the SCIENCE *of* BEING HUMAN

Also by Marty Jopson:

The Science of Everyday Life

The Science of Food

the SCIENCE *of* BEING HUMAN

WHY WE BEHAVE, THINK AND FEEL THE WAY WE DO

MARTY JOPSON

First published in Great Britain in 2019 by
Michael O'Mara Books Limited
9 Lion Yard
Tremadoc Road
London SW4 7NQ

A CIP catalogue record for this book is available from the British Library.

Papers used by Michael O'Mara Books Limited are natural, recyclable products made from wood grown in sustainable forests. The manufacturing processes conform to the environmental regulations of the country of origin.

ISBN: 978-1-78929-164-3 in hardback print format
ISBN: 978-1-78929-168-1 in ebook format

1 2 3 4 5 6 7 8 9 10

Typeset by Claire Cater

Printed and bound by CPI Group (UK) Ltd, Croydon, CR0 4YY

www.mombooks.com

For my father,
who took me to museums.

CONTENTS

INTRODUCTION

We all share many things in common with each other. If you are a fan of spicy food, board games, long walks in the countryside or early twentieth-century horror fiction then you have at least something specific in common with me. One of the few things that is certain, though, is that we all share being human. What does it actually mean to be human and what is the science behind it?

To find the answer to this I have taken an eclectic approach and poked about in branches of science you may not have expected. I came across some interesting nuggets and quite a bit of maths.

I have tried to look at being human from a number of angles, starting with where we came from and, as with previous books, I have aimed to keep it up to date with the latest scientific research. This has proven challenging, as we are in a golden age of new discoveries about our deep evolutionary past and the quirks of being human. The influence of bacteria in our lives, for example, and how they might even change our behaviour, is a very hot topic for scientists.

I also wanted to explore the role and place of humans in modern societies. Clearly, our world is no longer like the one in which our ancestors evolved, but we still manage to navigate it. The digital realm features in several sections of the book, as I consider how *Homo sapiens* that evolved as part of a hunter-gatherer community copes with being hooked up to the internet twenty-four hours a day. The interactions between technology and our own bodies are not always, despite what the adverts tell us, as simple as they sound.

Lastly, this book takes a look at an area often overlooked by popular science focusing on being human. To paraphrase, none of us is an island and we all live out our lives surrounded by other humans. Given the way the global population is climbing, we find ourselves in groups of increasingly larger numbers. How humans interact with each other in these groups breaks all the laws of physics that we assumed would apply. New rules and new paradigms have had to be invented to explain what happens when humans get together.

Which brings me onto the one big, take-home message I have truly appreciated in writing this book. Biology is messy. Physicists, engineers and, to some extent, chemists study the world with equations and the certainty of mathematics. Biological systems and, by extension, human beings, are gloriously, unnecessarily and inexplicably complicated and counter-intuitive. This is the reason I find some scientific topics more compelling than others, and to my mind there is nothing more fascinating than the science of being human.

JUST WHO DO YOU THINK YOU ARE?

A species by name

I am a member of the species *Homo sapiens*. This is not, I hope, too controversial a statement. Furthermore, I assume that you are also a member of *Homo sapiens*. It is the scientific way of saying you and I are both part of the human race. However, what does that really mean? It seems clear that we are all human and yet, once you begin to pick at this statement, it becomes a bit less certain.

The two words *Homo sapiens* form just the last part of the biological taxonomic system that allows a scientist to nail down precisely what type of animal, bird, reptile or plant is being talked about. The system was invented back in 1735 by one of the great scientists of the eighteenth century, a Swedish naturalist called Carl Linnaeus. Linnaeus's work was in Latin, which is why biological names still use this language. The system starts with the Kingdoms of Life. You may think this would be the easy bit, but unfortunately everything in the Linnaean system of classification has undergone, and is still undergoing, periodic change. We started out back in 1735 with just two Kingdoms of Life: animals and plants. That number has since grown, shrunk, grown again, shrunk back and is currently standing at seven recognized Kingdoms of Life. Starting with the tiny things, we have the kingdom of Bacteria and the kingdom of Archaea, which are a distinct and primitive form of bacteria. The next kingdom of Protozoa is made up of all the single-celled creatures like amoeba, larger and more complex than bacteria. The kingdom of Fungi is fairly straightforward, although far bigger than you may imagine, and plants are now divided into the kingdom of Chromista, where you find the algae and seaweeds, and the kingdom of Plantae, where trees and grass and such can be found. Lastly is our own place, in the kingdom of Animalia.

Once you have found your kingdom, you work down through phylum, class, order and family before finally reaching genus and species. In our own case, after the kingdom of Animalia we are in the phylum Chordata, all of whose members have some sort of spine and spinal cord. We are in the class of mammals or Mammalia and following this comes the self-explanatory order of Primates. Then, our taxonomic family is Hominidae, or great apes, that only includes the various orangutans, gorillas, chimpanzees, bonobos or dwarf chimpanzees and ourselves, humans.

Finally, we get onto the last two bits of our classification, our genus, *Homo*, and species, *sapiens*. Traditionally, so that they stand out in text, these two are always printed in italic and the genus is capitalized and sometimes abbreviated. The genus represents a very closely related group of different species. For example, *Panthera leo* is the lion and *Panthera tigris* the tiger. This double-barrelled naming system allows scientist to be precise and yet provide more information. Without knowing anything about *Panthera onca*, you can immediately gather that this species is probably some sort of big cat (it's the South and Central American jaguar). Similarly, if I tell you the domestic cat is *Felis catus*, you can see it is not that closely related to the lion as the jaguar.

But what does all of this mean in practical terms? Our genus *Homo* contains just one species at the moment, and that's us. In the past, there have been more species in the *Homo* genus – definitely another six and possibly another nine species on top of that – but they are all now extinct. What is a species and how do we draw the line between them? It turns out to be a much trickier problem than you may imagine. When Linnaeus first concocted the idea, it was primarily just an aid in identifying different types of plant when you were out in the field doing some botany. The basic concept was that a species breeds true. Which is to say that if the offspring of an organism are the same as the parent they qualify as a species. Even with this simple definition, other scientists argued with Linnaeus and with each other about how to identify a species. One important implication of this idea is that species are fixed and unchanging.

Then along comes Charles Darwin with all his ideas about evolution. Darwin was much perplexed by the nature of species and grapples with this in his seminal work *On the Origin of Species* (1859). In this book he wrote that he was 'much struck how entirely vague and arbitrary is the distinction between species and varieties'. The baseline for species changed to an understanding that two members of the same species and appropriate sexes should be able to breed and create progeny who themselves could breed and continue the species. But even Darwin saw this as problematic. According to his theory, species will evolve over vast time periods creating new species. At any one point in time the new species on the way to evolving is presumably still very much like the old species. At what point do they become separate?

Things became even more complex in 1942 with the work of Ernst Mayer, one of the leading evolutionary biologists of the twentieth century. He came up with the idea of biological species concepts and focused on not only the ability to reproduce but also geographical isolation. Since then, several dozen different biological species concepts have been put forward. Each one of these has its scientific adherents and the whole issue seems even less clear than it was when Linnaeus was thinking about it.

Take this example of a how biology seems to resist definition. The seabird genus of *Larus* gulls has a worldwide distribution with well over twenty different species. In 1925, the American ornithologist Jonathan Dwight realized that

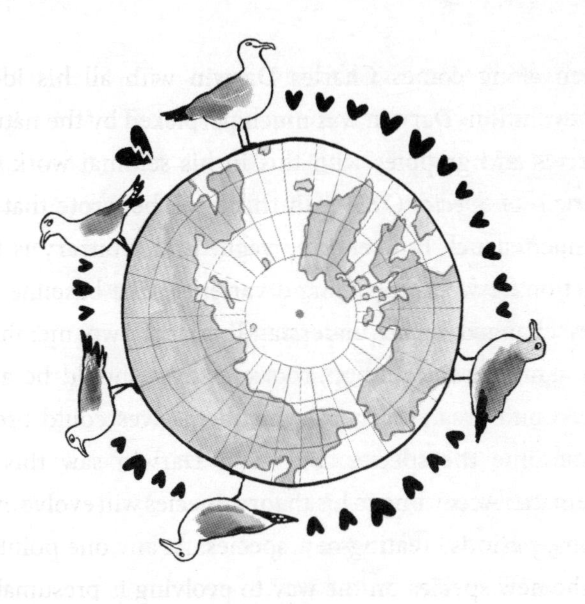

The *Larus* gull ring species

something peculiar was going on with the *Larus* gulls found all around the Arctic Circle. To spare you from too many Latin biological names I will stick to the common names for the birds studied by Dwight. Each of these species, it should be noted, while being of the same *Larus* genus, looks significantly different. The *Larus* gull I am most familiar with here in Great Britain is the European herring gull, which it turns out can breed and form hybrid non-sterile chicks with its western neighbour in North America, the American herring gull. In turn, the American herring gull breeds with the East Siberian herring gull, which breeds with Heuglin's gull, which breeds with the Siberian lesser black-backed gull. This last type of gull lives in the northern latitudes of Scandinavian countries and its territory rubs up against the eastern edge of the first gull I mentioned, the European herring gull in Great Britain. This is where the chain ends, as the Siberian lesser black-backed gull cannot breed with the European herring gull. The ring of interbreeding species is broken somewhere in the Norwegian and North seas. This peculiarity is known as a ring species and has been seen to occur very occasionally. If organism A can breed with organism B, even if they look different, then according to some biological species concepts they are the same species. But if organism B can also breed with organism C, then that makes all three organisms the same species, except that in a ring species organism C can't breed with organism A, which means they are different species. It all gets very complicated and our definition of species starts to break down. Especially as the latest genetic work on the *Larus* ring species shows that it may actually be two even more convoluted spaghetti tangles of connected interbreeding species, and not a ring after all.

What it means to be a species has become more nuanced as biological science has progressed. It has become apparent that the idea that each species is a distinct entity is merely a product of our own desires to categorise the organisms we find into neat boxes. Linnaeus created a system to help botanists and we have become stuck in his thought pattern ever since. All of which leads to paradoxical nonsense like ring species. You may assume that much of this only pertains to other organisms out there in the big wide world as after all, humans belong to a genus of only one species. Within the *Homo* genus there is only the *sapiens* species. But it was not always so.

A genus of one

There is a common illustration used to depict the evolution of the human species. It shows a line, usually just silhouettes, with a diminutive, crouching and unnamed chimpanzee-like ancestor on the left. Then follow, in progressive stages of increasingly erect postures, a line of figures striding towards the right. If the diagram is labelled, the figure next to the ape is an *Australopithecus* of some sort. After this we get to members of our own *Homo* genus, in turn: *Homo habilis*, *Homo erectus*, *Homo neanderthalensis* and, finally, *Homo sapiens*. The last two of these are usually depicted holding a spear or bow and arrow, to indicate tool use. It has become a popular image to copy and originally dates back to 1965 and a section of an American *Time-Life* publication called the Life Nature Library. In that book the image is entitled the 'March of Progress'. These days you tend to see it lampooned, with a final figure on the far right depicting modern man as either sat, crouched at a desk or clutching a beer and displaying a huge belly. There is a lot wrong with this picture, and it is not just the cheesy visual jokes it is used for.

The 'March of Progress' from 1965 is a powerful example of how an image can carry more weight than text. A close examination of the accompanying writing shows that the author was fully aware that the evolution of *Homo sapiens* was not a linear path. However, the power of such a simple image overwhelms any coincident writing.

For a start, it is always men that you see depicted. What it shows is the march of men to civilization and if they are holding something it is always a weapon. On top of that fairly obvious omission is what the great American evolutionary biologist Stephen Jay Gould called the implicit arrow of time. By placing the subsequent species of the *Homo* genus in a line, all seemingly moving from one to the next, you are making a not too subtle implication that each is in some way more advanced than the next. But evolution is not a process of increasing complexity, nor any other measure of superiority you can find. Evolution is blind, has no agenda and makes no choices. It is a random walk towards something that ends up being fit for purpose at the particular time and place. All of the species depicted in the 'March of Progress' were at some point perfectly adapted to their environment. As their environment changed, be that through change in climate, habitat or pressure from another species, that evolutionary fit to their niche may have diminished with the ultimate consequence of extinction. But none of this makes the other *Homo* species better or more evolved.

Aside from this fundamental and irksome error in understanding of elementary biological ideas, we now know that the 'March of Progress' is wrong in the details as well. The family tree of the *Homo* genus is far more complex and nuanced than we believed back in 1965. The earliest known species of human is probably *Homo habilis* that stood at a diminutive 1.3 m (4 feet 3 inches) and lived between about two million years and one and a half million years ago. This species then evolved into *Homo erectus*, the progenitor of all the later members of our genus. At least, that was the straightforward story told about twenty years ago. Since then a lot of other fossils have come to light that, rather than crystallizing the picture, seem to have muddied the water. We now know that *Homo erectus* and *Homo habilis* were alive at the same time and probably in the same areas. Then, between 1991 and 2005 a series of extraordinary fossils were found in a cave that lies halfway between the Black Sea and the Caspian Sea, in Dmanisi in Georgia. This series of five skulls showed a wide range of features that would appear to point to a number of different species being present, including *Homo habilis, Homo erectus* and even a couple of less well-known species, such as *Homo ergaster* and *Homo rudolfensis*. Yet the skulls were all found in the same place and are all of the same age, so the likelihood is that they are the same species. Which raises the possibility that all the early *Homo* species identified so far, often through just one or two fossils, are in fact the same species. You will also note that the Dmanisi fossils are a long way from Africa. The idea that humans evolved solely in the African Rift Valley and spread out from there is no longer seen as the case. While our origins do appear to be mostly African, it looks like early human species spread over a far wider range and intermingled far more.

Following the time of *Homo erectus* and all the other possible early species, we see the rise of genuinely distinct early humans from about 800,000 to 400,000 years ago. This is the age of *Homo heidelbergensis*, a species named after a fossil jaw discovered in 1907 by Daniel Hartmann, a workman at an open-cast sand mine just outside Heidelberg in Germany. Hartmann reported his find to a local professor of anthropology called Otto Schoetensack, who went on to name the species. Specimens of *Homo heidelbergensis* have since been found in locations from South Africa all the way up the east coast of that continent and into Europe where it has been found it Italy, Greece, Spain, Germany, France and even across into the UK. What makes *Homo heidelbergensis* particularly interesting is that, according to the current theory, it led to the evolution of three further species of the *Homo* genus. One of these is us, *Homo sapiens*, although there is too big a gap in the fossil record to really nail down this evolutionary link. The second species is known only as the Denisovans, but more about them later. However, the last of these species that traces similar ancestry is familiar to us all and their name is used as a synonym for brutes and brutish behaviour. Neanderthal man was named for a partial fossil skull found in a limestone quarry in 1856 in the eponymous Neandertal valley in Germany. Although this specimen gave the species its name, *Homo neanderthalensis*, it was not the first Neanderthal specimen to be found. That honour goes to a crushed child's skull found in Belgium in 1829 and a fine partial skull was also discovered on the Rock of Gibraltar in 1848. You may have noticed that there is a discrepancy in the naming of this species, Neanderthal man found in the Neandertal valley. The loss of the letter h from thal came about in 1901 when the German language dictionary published by Konrad Duden was made the official arbiter of spelling. The regional spelling of thal, meaning valley was standardized, and the h was lost, but by then the Neanderthal species' spelling was fossilized.

We now know that Neanderthals ranged widely across southern and central Europe and right over to the far side of Kazakhstan and the edges of Mongolia. They definitely used tools, as numerous flint flakes of theirs have been found and, in one case, wooden spears. However, until 2018 the jury was out as to their artistic ability. Before this, only a few items and scraped marks had been found that indicated any more complex social culture. Then, in three separate caves spread around Spain, a collection of cave paintings showing red lines, dots, a ladder pattern and a hand stencil were dated at 64,000 years ago. This makes them the oldest known cave paintings and places the artists well before *Homo sapiens* were in the area: the only culprit could have been Neanderthals. These were clearly not the hulking caveman brutes of pulp films as they showed a cultural sophistication at least the equal of *Homo sapiens*.

While modern *Homo sapiens* only reached Europe about 50,000 years ago, we have managed to move the dates of our own species much further back. A find at Jebel Irhoud near the Atlantic coast of Morocco has been identified as a modern human and in 2017 was dated at 315,000 years old. This pushes our own species much further back in time than previously thought, at least on the African continent. In addition, this deepening of our history starts to correlate with some of the genetic evidence of human evolution.

The evolutionary history of the human species is far from being a conveniently linear 'March to Progress'. As we saw in the last chapter, it is hard enough with related living species to work out where one ends and the other begins. The task of working out our own lineage when all we have to go on is fossils is made much harder. The early *Homo habilis* may have been a separate species or one and the same as *Homo erectus*. Furthermore, it looks like *Homo sapiens* has been around for a very long time, although not sequentially but simultaneously with other *Homo* species. Which raises the interesting and potentially knotty problem of what happened when humans met Neanderthals?

When human met Neanderthal

In 1977, Frederick Sanger and a team of scientists in Cambridge, Great Britain, published the entire genetic sequence or genome of a virus called lambda-X 174. It was a groundbreaking moment in science as this was the first complete genome of any organism to be determined. Since the work of Watson, Crick and Franklin, in 1953, we knew that deoxyribonucleic acid, or DNA, is the genetic material in all known living things and many viruses. The long strands of DNA with their four-letter codes represent the instructions for making an organism. For this first genome to be sequenced, Sanger chose the subject carefully. The virus lambda-X 174 has a tiny genome consisting of only 5,386 of DNA's four-letter code. Other genomes then followed as laboratories around the world started to sequence viruses, bacteria, yeast and tiny roundworms. However, the prize was the human genome and in 1984 an international group of scientists began to plan for this mammoth task. It was a formidable undertaking, as the human genome contains over 3 billion of DNA's letters. The Human Genome Project started in 1990, cost about US $3 billion and when it was declared complete on 14 April 2003, was and still is the largest biological project ever carried out. At the time of writing, you can have your entire genome sequenced in about an hour, using a handheld device for around a thousand US dollars. The advance in technology alone is mind-blowing, but what this has enabled us to do is equally remarkable.

By comparing human genomes from around the world and from different cultural heritages it is possible to build a genetic map of the evolution of the human species. Since we know and can make assumptions about the rate at which random variations build up in a genome, you can use this information and the number of differences between two genomes to get an idea of when those organisms began to take different evolutionary paths and diverge. Based on this data and the fossil record we can draw up the picture of the path of evolution taken by species of the *Homo* genus. But could this approach be used to investigate the earlier, now extinct human species?

Initial attempts to examine ancient genetics looked at Neanderthals, the human species that had most recently become extinct about 40,000 years ago. To make life easier for the researchers they looked at the tiny loop of DNA found inside sub-cellular organs called mitochondria. It was still a huge task, but when this DNA had been sequenced and compared to modern human mitochondrial DNA it indicated a date of separation between the two species of about half a million years ago. Then in 2006 the international Neanderthal Genome Project began, based in Leipzig in Germany. The scientists extracted DNA primarily from the long leg bones of three female Neanderthals found in a cave in Croatia, dated at 38,000 years old. It took them four years and when in 2010 they finally released the results there were incendiary implications. The work on mitochondrial DNA had not shown any mixing of the genetic material between Neanderthals and humans, but with the much larger and more complete picture from the whole genome it looked like there had been a blending or admixture of human and Neanderthal. The genetics allowed even more detail to be elucidated. This admixture took place about 50,000 years ago in the far east of the Mediterranean, around what is now Syria, Israel, Lebanon and Jordan. Which is the delicate way to say that members of the human species had sex and produced offspring with their Neanderthal neighbours, and this must have been a fairly commonplace occurrence as for non-African humans between 1 and 4 per cent of our DNA has Neanderthal origins.

This is not the only instance of interbreeding that cropped up. Recall the Denisovan human species I mentioned in the last chapter? It looks like humans and Neanderthals bred with

Denisovans as well. The only fossil evidence we have for this species are a scant few fragments found in a cave in 2008 in the south-west of Siberia in Russia. So far archaeologists have unearthed a total of three Denisovan teeth, a tiny finger bone and a 25 mm (1 inch) sliver of arm or possibly leg bone. It is the paucity of these finds that explains why the Denisovans have not been given a proper scientific name. Despite the rarity, it was decided that the finger bone should be used for genetic analysis. A genome was produced and it became clear that the owner of the finger bone was neither Neanderthal nor modern human. We don't know what Denisovans looked like, how tall they were or any other details, but we do know that they interbred with Neanderthals and with humans on many occasions. When the Denisovan genome was compared to different people around the world, those living in Melanesia, the islands of Papua New Guinea out to Fiji, had up to 6 per cent Denisovan DNA. From this work it has been possible to map the extent of this mysterious early human species. They ranged all across Asia and down though Polynesia and into Australia, where Australian Aborigines have a small component of Denisovan in their genome.

This web of interbreeding was shown at its most obvious when the genome of an ancient human called Denny was sequenced in 2012. Recall that the only fossil remains of Denisovans are three teeth, the finger bone that yielded the initial DNA sequence and a leg or arm bone fragment. When this leg or arm bone fragment was analyzed it was found to come from a girl of about thirteen years old. Remarkably, it turns out that her parents were of different species. We can say with some certainty that her father was a Denisovan and her mother a Neanderthal. Not only that, but we know that the cave in Denisova where these five precious fragments of our past were unearthed also contained *Homo sapiens* remains. It would seem likely that three species of human – Denisovans, Neanderthals and *Homo sapiens* – all lived in the area at the same time and quite possibly in the same caves with each other.

There is a sixth fossil remain that may be of Denisovan origin, found in 1980 by a monk in the Baishiya Karst Cave in Tibet, a long way from Denisova. While DNA analysis failed to reveal anything, in 2019 a German team re-examined the fossil and matched collagen protein found on it to the Denisovan gene for collagen protein.

There are implications to all this cross-breeding that took place, some positive and some negative. Many Tibetans appear to have a Denisovan version of the EPAS1 gene that is associated with adaptation to living at high altitude, which they certainly do. This alternate Denisovan version of EPAS1 was clearly of benefit to those living at such elevated heights and has remained prevalent in their genome, whereas other ethnic groups have lost the gene by a process known as genetic drift. While having the variant gene at low altitude gives no benefit, it equally conveys no hindrance. Evolution does not select against it but it does fade from the lowland population by random chance.

A whole range of diseases appear to be linked to the interbreeding exploits of our species, including Crohn's disease, some forms of lupus and even Type 2 diabetes. A common factor is that these are all autoimmune diseases and the cause is an important group of genes known as human leukocyte antigens, or HLAs, that help us identify which cells in our body belong to us and which represent invading pathogens. A good half of the HLA genes can be traced back to either Denisovan or Neanderthal ancestors. When our body fails to properly recognize the HLA gene products, the result is an autoimmune disease. Case in point is the unpleasant Behçet's disease, an inflammatory disorder that affects the whole body and is caused by a Neanderthal version of an HLA gene.

Not all our inherited and interbred genes are bad for us though, far from it, and it would appear that it was our predilection for interbreeding that resulted in some of the most significant evolution of our species in recent time.

The ever-evolving human

There is an argument that says that modern humans are no longer subject to evolution by the process of natural selection. According to this idea, we are now beyond the biological processes that created not only *Homo sapiens* but also led to all the biological complexity and diversity we see around us. We have broken free of the biological hold on our evolution. On the surface it may appear that this is true, in so much as we are now able to change our environment to suit our biology, rather than the other way around, and alleviate the pressures of natural selection evolving us to fit a specific environmental niche. However, our current understanding would indicate that the pace of human evolution is anything but static.

The process of evolution as outlined by Darwin back in 1859 is not, as is often indicated, a one-step process. There are two parts to evolution, something that Darwin makes clear not only in the text of his seminal work but in the title: *On the Origin of Species by Means of Natural Selection*. The first part of the title refers to the origin of species, or for brevity you could say evolution. This is the process whereby one species gives rise to another. A prerequisite for evolution to take place is variation within the population. Long before Darwin, scientists had hypothesized that species were not fixed and could change gradually into other forms. For this to happen you need a certain amount of difference between individuals on a generation-by-generation basis. After all, you do not look exactly like either of your parents, nor are you an exact mixture of the two. Instead, while you may have some resemblance to your parents, your features are uniquely yours. There is plenty of variation within the human population, even if you restrict yourself to physical appearance. Dig into human genetics and the possibility for variation increases manyfold.

Contrary to popular belief, Darwin avoided the subject of human evolution in *On the Origin of Species* (1839). *The Descent of Man* (1871), which immediately sold out twice, included it. His friend Joseph Hooker commented, 'I hear that Ladies think it delightful reading, but that it does not do to talk about it, which no doubt promotes the sale.'

The second part of Darwin's idea is about natural selection, the process by which evolution takes place. At the heart of natural selection is that you must be able to pass on any variation that gives you a selective advantage. To put it bluntly, if through random variation you are born different to the rest of humanity and have an amazing and beneficial mutation it won't make a scrap of difference if you do not have any children and thus don't pass on this mutation. What is more, natural selection does not necessarily work in favour of an organism. Natural selection is blind and has no forward plan. Take the human eye as an example. While it is a marvel of organic design it has one major defect that no engineer would include. The optical nerve fibres that connect to the individual parts of your retina, the light-sensitive bit of the eye, do not approach the detection cells from the back, but from inside the eye ball. It's the equivalent of trying to wire up the screen of your television by running all the cables in front of the screen, obscuring the view, rather than hiding the wiring behind the screen. It would be a crazy way to plan a design, but this is how your eye works. The reason it is this way is just a quirk of biology, an accident of unplanned evolution and a consequence of blind natural selection.

Human beings are still subject to evolution by natural selection. We still have variation between individuals and our environment still exerts pressures on us although these days subtler ones than on our early hominid ancestors. The brutal reality of nature is more distant when you can just make a trip to the supermarket for food. Instead there are the pressures of modern civilization, information technology and new psychological challenges imposed by our culture. Taken together, variation with selective pressures drives change. But that does not mean that *Homo sapiens* is going to suddenly evolve into a new species any time soon. First of all, recall the problems of how to define what constitutes a species in the first place as discussed in an earlier chapter. While our genetics tell us that we are different from humans of 300,000 years ago, our skin colouration has diversified for a start, it is not enough for us to be classified as a different species.

Also, human evolution is a very slow process, at least on the scale of our lifetimes. A study that looked at the evolution of new species and specifically changes in size between species estimated that it took at least a million years to really cement the change and for the species to be distinct. We humans have been around for only a third of that time.

Yet there are some small but definite examples of evolution that we can see in action in the modern human. Take milk, for example: a wonderful food for an infant mammal. It has everything a growing body could need and contains a full complement of essential protein- and fat-building blocks. It's loaded with vital minerals and vitamins such as calcium and vitamin B. On top of this, it is packed with energy in the form of sugar. Except, nearly all adult mammals cannot digest it and its consumption will result in stomach pains, vomiting, diarrhoea and flatulence. About two thirds of the global human adult population cannot stomach milk. The problem is lactose, the sugar that all mammalian milk contains. Lactose is a disaccharide, because each molecule is made up of two smaller monosaccharide sugars stuck together. In the case of lactose, it is a unit of glucose stuck to a unit of galactose, whereas sucrose, or table sugar, is made of glucose linked to fructose. However, for mammals to be able to digest lactose we need to cut the molecule in half, separating the glucose from the galactose. Fortunately for infant mammals, including baby humans, we all have a gene encoding for an enzyme called lactase. When we are born this gene is active and we have no problem digesting all the milk sugars. However, since mammals do not encounter milk as a food after they have weaned, mammals evolved to shut down the production of this enzyme when it was no longer needed. Which is why when two thirds of the world's adult human population drink milk they suffer the aforementioned unpleasant effects. The lactose remains in the gut and provides a bumper meal for normal intestinal bacteria, leading to fermentation inside your bowels. It is known as lactose intolerance, but one third of us, myself included, do not suffer from it. Instead we have what is known as lactase persistence.

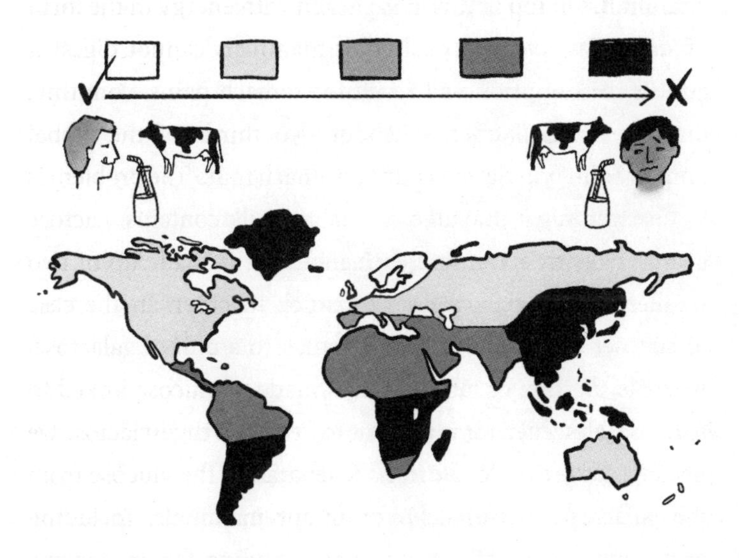

Distribution of lactose intolerance

In those with lactase persistence, the gene that encodes for the lactase enzyme does not switch off after weaning. Lactase continues to be produced in the gut and milk can remain a viable food. If you look at the distribution of lactase persistence around the globe it ranges from nearly 100 per cent in Irish people, more than 90 per cent in the rest of Northern Europe, right down to 10 per cent in China, less than 5 per cent in Native Americans and it is completely unknown in the Bantu people of South Africa. The genetic mutation that changed the lactase gene appears to have only arrived in our genome in the last 9,000 years. The current theory is that during what has become known as the Neolithic revolution, when we switched from hunting and gathering to farming, we started looking at our livestock as a walking source of nutrition in the form of milk. The ability to digest milk became an advantage and this genetic variant spread in the population, evolving the people.

The beauty of the lactase / lactose example of evolution is that you can see it happening in populations right now. South American people have a very low incidence of lactase persistence and milk is not a common food for adults. Yet in Chile, between the Atacama Desert and the country's fertile valleys, there is a population with a very high lactase persistence relative to their neighbours. The Chileans living in this hard, dry landscape grow crops on small plots of land and herd goats – lots and lots of goats. Herding goats has only been practised in the area for a few hundred years, as goats were introduced by European colonizers. Naturally, the herders of goats make use of goats' milk, which contains lactose. It is an evolutionary blink of an eye since the goat became popular there and yet more than half the population is now lactase persistent. What is more, studies of the population show that those with the ability to digest milk are better nourished. The assumption has been made that this selective advantage is driving the new lactase persistence gene to flow through the population with each successive generation. The Chilean goat farmers are evolving.

The acquisition of the ability to digest milk is a fairly straightforward example of evolution in action. There is a single gene involved and it is easy to test for lactase persistence in the population. However, do not be fooled by its simplicity, as what it demonstrates has profound implications for humans. No matter how much you think our sophisticated culture has separated us from other animals, the forces of natural selection break through and we are all still evolving.

An atlas for the human body

What are you made of? It is a fundamental question that has a range of different answers. To a physicist the answer may involve protons, neutrons, electrons and, if you persist, quarks, leptons and bosons. A chemist would possibly back up a little and tell you we are made of atoms of different elements, mostly carbon, hydrogen and oxygen, organized in complex molecules. However, for a biologist the answer is cells. All living organisms are made up of at least one cell.

In 1665 a man named Robert Hooke published a remarkable and beautiful book called *Micrographia*. At the time, Hooke was the Curator of Experiments for the recently formed, London-based Royal Society. The Royal Society still exists, just up the road from Buckingham Palace, and it is now the oldest scientific institution in the world. Hooke's job was to set up and perform experiments for the members of the Royal Society, either of his own devising or at the behest of the members themselves. It was a big step up for Hooke who, unlike the members of the Royal Society, was not wealthy or aristocratic. Through sheer determination he had become an educated man in the new study of natural philosophy, or as we would call it today, science. Armed with an inquisitive mind and a remit to try out pretty much anything he wanted, he got hold of a microscope and began observing everyday objects. The book he produced is filled with extraordinary illustrations on large fold-out sheets showing ants, spiders, the surface of a nettle leaf, samples of cloth, moulds and the surface of the moon, which presumably he did not observe through his microscope. Accompanying each image is a careful description of what he saw and the implications of the things he noticed. Given its age, the book is surprisingly readable and full of interesting asides. Possibly the most significant contribution, though, can be found in observation eighteen. In this chapter, Hooke describes how he carefully sliced a wafer from a piece of cork and looked at it under his microscope. He explained that he saw 'pores, or cells' that were 'not very deep, but consisted of a great many little boxes'. This was the first time that the tiny components of all living things were described and given the name of cells. The hand-drawn illustration that went with the text, while not as beautiful as Hooke's ant or as detailed as his notorious picture of the flea, is one of the most significant images in the history of biology. It shows a series of almost rectangular grey shapes surrounded by white walls and arranged in closely packed and ordered lines. It is not surprising that Hooke saw these tiny boxes and drew upon a word normally associated with rows of small monastic rooms. Within his description of the cork cells he goes on to realize that many other plant materials he observed seem to possess similar structures and in some cases they are filled with a fluid or 'the juices of vegetables', as he put it. It is a remarkable piece of work slightly spoiled when he incorrectly draws the conclusion that cork is a fungal growth on the surface of the cork oak tree and not, as we now know, a thickening of the tree's bark as an adaptation to survive forest fires. Despite this error, the first observation of cells opened the way for modern biology and our understanding of how human beings work.

Given that we have known about cells now for over 350 years, you may be surprised to find that we still don't know all the different types of cells that make up a human being. We do know that the human body contains about 37 trillion cells – which is 37 followed by twelve zeros. We are aware of thousands of cell types but it has become apparent that there are many more. There were obvious cell types like red blood cells, nerve cells, white blood cells, skin cells, sperm cells, egg cells and so on, but the closer biologists looked, the more subtle the variations appeared.

Just a few types of blood cell

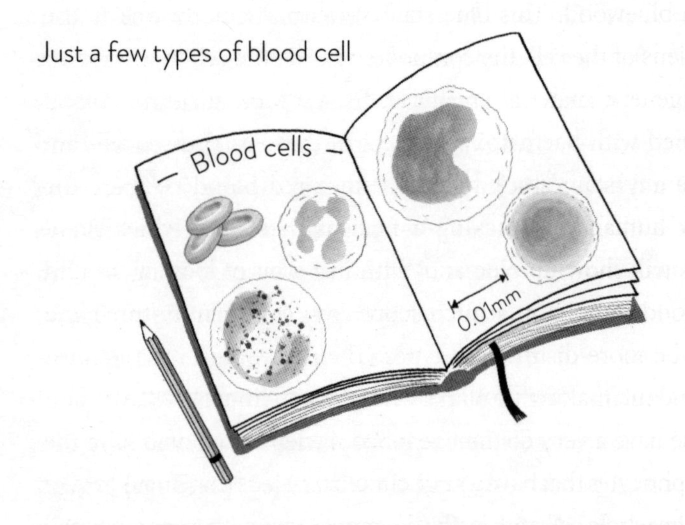

Take blood cells as an example. There are two main types of cells within blood: red blood cells and white blood cells. People had seen red blood cells about the same time as Hooke was publishing *Micrographia*, but it was another 175 years before white blood cells were discovered. The problem, and it is a common problem in microscopy, is that white blood cells are essentially see-through and invisible. It was only with the revelation that dyes could be added to colour cells under microscopic observation, making the invisible visible, that cell biology blossomed and, in the case of blood, we discovered not one but dozens of types of white blood cells. By carefully selecting your microscopic dyes, you can stain some components of the inside of the cells but not others. Take for example haematoxylin, a deep blue compound extracted from the heartwood of a central American tree known appropriately as a bluewood. This blue stain sticks particularly well to the nucleus of the cell, the component inside the cell that contains the genetic material and DNA. If you look at human blood stained with haematoxylin, most of the cells you see will not have any stain stuck to them, since red blood cells are the only human cells lacking a nucleus. However, white blood cells will show up blue and, with just a bit of looking around a blood sample on a microscope, you can identify three and maybe more distinct cell types. The neutrophils are the most abundant, making up 60 per cent of your white blood cells, and these have a very distinctive lobed nucleus. Then you have the lymphocytes that have a huge circular nucleus that seems to take up the whole cell and, lastly, the monocytes with a bean-shaped nucleus. With practice and ideally another stain you can also spot basophils and eosinophils, which brings the white blood cell type tally up to five. Each of these cell types has a completely different function. The neutrophils prowl the blood looking for bacteria to engulf and destroy, while the monocytes, when activated, undergo a remarkable transformation turning them into another type of cell altogether: hunter-killer macrophage cells that escape the blood and roam inside the tissues of your body. Once science began to really unpick what white blood cells did and how they worked to create our immune response, the humble and dull-looking lymphocytes turned out to be at the heart of the system. Not only that, there are three distinct flavours of lymphocyte: the B-cells make antibodies, the T-cells identify foreign agents in our bodies and the natural killer cells seek out and destroy our own cells that have become infected with viruses. This only skims the surface of the immune system's cell types. There are currently thought to be over twenty types of cell involved in this one system alone. What is more, the deeper our understanding of the immune system, the more cell types are discovered. All of which explains why one of the most exciting biological projects running at the moment is the international Human Cell Atlas.

The first organism to have a complete cellular map created for it was the tiny, virtually see-through, one-millimetre-long roundworm that lives in soil, called *Caenorhabditis elegans*. This tiny worm only contains 959 cells and the fate and origin of every single one of them was determined in 1977 by scientists working in Cambridge, UK.

Understanding how a whole organism functions relies ultimately on how the cells within that organism work together. This is fairly straightforward if you are studying a single-celled amoeba or a tiny worm with only about 3,000 cells. But if the biology you want to really understand is our own, then you have the aforementioned 37 trillion cells to cope with. To create a model for how all these cells work together to give us a human you need to understand what the different cells can do, where they are in the body and then you can start to wonder how they interact. Up until now biologists have been plugging away at the problem, finding new cell types, working out what they do and slotting them into our understanding. In 2016, two biological researchers got together and made a plan: Sarah Teichmann, working in Cambridge in the UK and Aviv Regev, who also works in Cambridge, but confusingly in Cambridge, Massachusetts, in the USA. Teichmann and Regev had both been looking at ways to uncover the pattern of genes within a cell that were actively being used and those that were dormant. It turns out that in cells that are not really doing much or growing rapidly, only 2,000–3,000 of the more than 20,000 genes in your DNA will be active. The rest of the genes will be turned off and inactive. By looking at which genes are turned on, it is possible to identify from just the genetics what cell type any one cell belongs to. New techniques were being developed that allowed this analysis to be carried out on a single cell, with the sample plucked directly from living tissue. What Teichmann and Regev realized was that the state of this technology was now at a point where they could propose something as radical as the Human Genome Project that set out to map all the DNA in a human. The Human Genome Project was a massive undertaking that started in 1990 and took sixteen years to complete. The DNA sequence of the final chromosome was published in 2006, after many thousands of scientists had toiled on the project, spending billions of US dollars. The results changed our understanding of biology and the Human Cell Atlas has the potential to do just the same.

The project to create the Human Cell Atlas is now well underway, with over a thousand scientists working in 584 different institutes from 55 countries as diverse as Ecuador, Nigeria and Russia. In April 2018 the project released its first batch of information and detailed over 500,000 cell types. This data is already beginning to shed new light on our understanding of human biology. For example, Sam Behjati, who works with Teichmann, identified that the cells in a common form of childhood kidney cancer, called Wilms' tumour, were in fact just foetal kidney cells that had failed to develop normally and instead divided aberrantly to create a tumour. The implication of this is that rather than treating Wilms' tumour with traditional chemotherapy, if the correct protein signals could be worked out, the cells could be converted to mature kidney cells instead. Another breakthrough came when a completely new type of cell, dubbed the ionocyte, was found in the walls of the lung. What makes the ionocyte interesting is that it looks like it may be causally involved in the relatively common disease cystic fibrosis. Previously, a different cell type was thought to be involved and drug therapies were developed to target it. With the discovery of ionocytes, more effective treatments for cystic fibrosis will hopefully be in the pipeline. Focusing on these few early examples, though, may be missing the point of the Human Cell Atlas, the aim of which is to provide a foundation for exploring all of human biology.

The biological meaning of the word cell would never have come about if Robert Hooke had realized that all of the specimens he was observing under his microscope were composed of these building blocks of life. It was only luck that he made his discoveries about how large organisms are put together by observing a slice of cork. The limitations of the technology he had to use prevented him from applying his ideas elsewhere and thus he chose to describe the microscopic boxes with reference to the living arrangements of monks. Most cells being identified for the Human Cell Atlas bear very little resemblance to cork cells, let alone monastic ones. I consider myself very fortunate to have had the opportunity to visit the Royal Society in London where Hooke worked and in its wonderful, gilded library I was allowed to take out and go through Hooke's own copy of *Micrographia*. The pictures of insects, flowers and fungal spores are gorgeous and intricately detailed, but the image that left the greatest impact on me was the simple microscopic examination of cork.

SOME UNCOMFORTABLE BIOLOGICAL TRUTHS

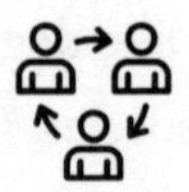

Are you domesticated yet?

One of the most remarkable things about humans is our ability to get on with each other. Now, you may be thinking that this is an unwarranted claim. After all, every time you look at the news there are endless reports of humans being violent and horrible to other humans. Death and war seem to constantly stalk the pages of newspapers and television reports. However, scientists and historians have for some time questioned this as a knee-jerk response to a diet fed to us by the popular media. The most eloquent counterargument to this feeling of increasing violence was made by Steven Pinker, the acclaimed cognitive psychologist based at Harvard University in the USA. Pinker analyzed deaths across history and came to the conclusion that, rather than living in an age of war and violence, we are in the midst of an era of unprecedented peace. This is all the more counter-intuitive as the population of the world is ever increasing. It looks like humans really are sociable creatures, quite unlike our closest cousins. Large groups of apes, especially chimpanzees, tend to become violent and unstable. We humans are also particularly good at non-violent reactions to strangers, unlike chimps where encounters between two groups usually lead to violence. It seems like humans have become much less aggressive. Ironically, when this has happened in other species like dogs, cats, sheep or horses the usual cause is the human species. One of the key elements of domestication of a species is to make it less aggressive and more inclined to interact with humans non-violently. So, have we domesticated and tamed ourselves?

The most influential scientific study on domestication took place on a farm just outside Novosibirsk in Siberia. In 1959 the Russian zoologist Dmitri Belyaev, inspired by the new science of genetics and his previous work in the Soviet Department of Fur Animal Breeding, decided to attempt to domesticate the silver fox. It should be noted that this was a brave thing to do. During Stalin's leadership of the Soviet Union, one man, the agronomist Trofim Lysenko, laid down the law when it came to biology. Lysenko rejected Darwinian evolution and claimed that the science of genetics was anti-communist. As a result, thousands of biologists were either executed or sent to labour camps. Belyaev started his work after Stalin had died: the Soviet Union had entered the Khrushchev era, but while genetics was no longer banned it was still a risky business. Lysenko's influence had waned, but in 1959 the fox experiment came close to being shut down before it had a chance to get going. In an attempt to regain his political power, Lysenko formed a committee that condemned the work of the institute that Belyaev worked for. Then later that same year, returning from a visit with Mao Tse-Tung in China, the Soviet leader Nikita Khrushchev decided to look into the situation for himself and paid a visit to Novosibirsk. At the end of the visit Khrushchev was apparently ready to shut the whole institute down. According to reports from staff, it was only the presence and presumably the influence of Khrushchev's daughter, Rada, a journalist and trained biologist, that he relented and only sacked the institute's director. Belyaev, then the deputy director, was promoted and his one-year-old fox experiment could carry on.

While it is the domestication of animals that gets the most interest, the production of domesticated plants has had a more profound effect on the human species. Without uniform, high-yielding varieties of the major crop plants such as wheat, maize and rice, the agricultural revolution that enabled the development of larger communities and technology would never have happened.

While Belyaev instigated the experiment, it was his research assistant Lyudmila Trut that did all the work with the foxes. Trut set up a breeding programme and subjected each pup at the age of just one month to a simple test. Trut would attempt to pet and hold the animal while she fed it. The test was repeated on a monthly basis over the first half year of the pup's life with no other significant human contact between the tests. At the end of this process the pup was rated on a tameness scale and only those in the top fifth of each generation were allowed to breed. Within just six years and six generations of artificial selection for tameness there were foxes that were happy to be handled by humans, wagged their tails, whined when they were left alone and licked the hands of experimenters. By 1979 and the twentieth generation of foxes, about a third of the fox pups when tested fell into the most tame category. In 2009, that figure had more than doubled, showing that it took just fifty generations to well and truly domesticate the silver fox. In those fifty years the changes were not only social and behavioural; the foxes had also changed physically. Their skulls had become less heavy set and more delicate, their snouts had shortened, their fur changed, becoming mottled or even red, and their canine teeth reduced in size. This had been expected, based on studies of fossils and observations of other domesticated animals. It seems that domestication comes with a package of not only increased sociability but physical changes to the body, especially the skull. Which is exactly what appears to have happened to us.

A study of ancient human fossils carried out by scientists from the USA found that at some point between 90,000 and 80,000 years ago our skulls changed a bit. The heavy brow ridge above our eyes and the overall height of our skulls seem to have diminished. Furthermore, when compared to *Homo erectus* and *Homo heidelbergensis,* our jaws have become less pronounced and drawn back under our skulls. Our snouts have shortened and our canines reduced. These are all characteristics associated with domestication seen in not only fossil records, but also Belyaev and Trut's foxes. It is a reasonable hypothesis to suggest that, if our skulls display the marks of domestication, then the reason for our extreme sociability is this domestication. This makes a certain amount of evolutionary sense, as for evolution by natural selection to take place you need to be able to pass on your genes through procreation. As the population of humans rose and communities grew, those of us who were more able to deal with the pressures of interacting with lots of other humans would be more likely to successfully negotiate the social complexities of finding a mate, having children and then rearing those children to adulthood. Being more social and less prone to aggression towards strangers would be an evolutionary advantage. The very fact that humans need to interact with each other in large groups creates the same selection pressures that Belyaev and Trut imposed artificially on their foxes.

While the superficial evidence seems to indicate that humans, like wolves, horses and cattle, have undergone a process of domestication, this just tells us what has happened, not how. One of the first changes that Belyaev and Trut found in their friendly foxes was a reduction in the levels of adrenaline in the blood of the pups. Further work has shown that reduction in testosterone may underlie some of the social and physical changes that take place. However, the most interesting new findings have drawn a correlation between the common features of domestication and the fate of a type of cell found in the very early embryo, the neural crest cell.

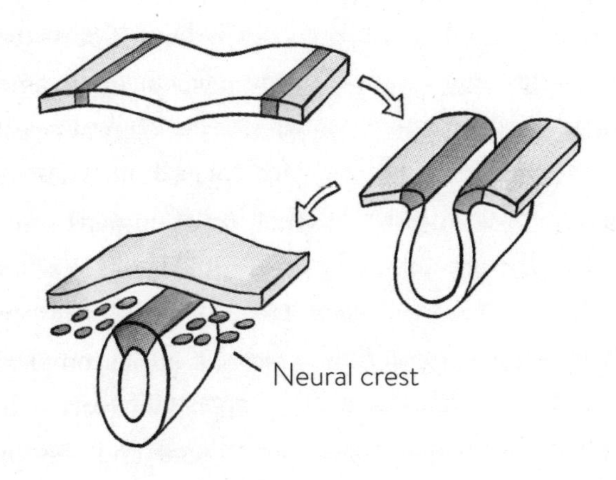

Formation of the neural tube and neural crest cells

One the first things that happens during the development of a human or any mammalian embryo is the creation of the neural tube. About eight days after fertilization, a human embryo is just a hollow ball of cells. At this point, the developing surface of the embryo folds in on itself to create an internal tube that is then pinched off. This tube goes on to develop into the spinal cord, the brain and the nervous system of the growing person. The cells that help drive the formation of the neural tube and are left behind once it has been completed, are neural crest cells. Those left-behind cells go on to develop into a whole range of different tissues and structures within the body, from the bones of the head to pigment cells in the skin, the adrenal glands and even teeth. It is a collection of disparate features that seems to map rather well onto those seen as part of domestication syndrome. In Belyaev and Trut's experiment, the foxes' skulls changed shape, the teeth were modified, the colour of the skin was affected and the drop in adrenaline was presumably down to changes in the adrenal glands, where it is produced. This correlation is tantalizing and may indicate that a reduction in neural crest cells may be the underlying cause of domestication. Currently, it is just a hypothesis and at the time of writing there is no conclusive evidence to put a causal link between neural crest cells and domestication. If that evidence does appear it will be a window into how it is humans became so very sociable, domesticated and tame.

The complexity of colour

Society in the twenty-first century is still attempting to come to grips with the colour of your skin. It is without doubt one of the most emotive issues in our modern culture. So, it should come as no surprise that how we ended up with such a wonderful and varied palette of skin colour has been extensively studied. Given that this is biology, and biology is rarely a simple story, it is also no surprise that the more we discover about the science behind skin colour, the more confusing it seems to become. While the 'why' of skin pigmentation is still under debate, the 'when' has been relatively conclusively worked out.

All of the work on the dating of skin colour has only happened since the revolution in our ability to extract and analyze ancient DNA. Skin and hair, or any trace of these structures, do not leave a fossil record. So, there is no way from dry fossils alone to decide on the skin colour of an ancient prehuman or even human. Skin colour is dictated by the presence or absence of a chemical called melanin that comes in a variety of different forms. Some of these have nothing to do with colouration, but the most common varieties are a brown and a black pigment. Melanin is manufactured in melanocyte cells sitting right at the base of the top layer of the skin, the epidermis. These cells have a spidery, starfish shape, with long protrusions extending off into the rest of the epidermis. Inside the melanocytes, melanin is packaged up into little bundles and sent right out to the end of these long protrusions, where it is gently nibbled away by the surrounding skin cells. Consequently, the dark melanin pigment is spread not just in the melanocytes but also the surrounding skin. The colour of your skin depends on how much melanin your melanocytes make and this is determined by the melanocyte-stimulating hormone receptor or MSHR. The MSHR protein sits on the surface of melanocytes and when it gets a hormonal signal from the pituitary gland it turns on melanin production. How much MSHR you have depends on how active your gene for MSHR is and this is ultimately the reason why some people have dark skin and others light skin. Once this had been worked out geneticists started looking for the presence of this gene in ancient DNA. When they looked at human genomes from very old human fossils all of them had the MSHR gene and thus were likely to have been dark-skinned. In fact, when geneticists dug back through our evolutionary history it appears that dark skin first evolved at least 1.2 million years ago, before the human species existed and coincident with the loss of much of the hair from our bodies (see page 137).

So, it seems that when *Homo sapiens* first arose as a distinct species, we had dark skin. The evolution of light or fair skin takes place much, much later and by the looks of it has happened twice in different places. For Europeans it seems to have taken place only about 6,000 years ago. Recall that our species first walked into Europe about 40,000 years ago. So, for 85 per cent of our occupation of Europe, humanity was dark-skinned. The researchers that discovered this identified a change in the genetics of early Europeans that, rather than being part of the MSHR gene, was a different mutation further along the line of melanin production. The lighter colouration of East Asian people has its roots in yet another genetic change, although at the time of writing the date of this change has not been found.

While it is now possible to get a fairly accurate date on the ancient appearance and then relatively recent disappearance of dark skin for Europeans at least, why this happened and what evolutionary forces drove it are currently up for discussion.

> Hair colour is also determined by the proportion of three kinds of melanin produced by the stem cells in the hair follicle (see page 138): black eumelanin, brown eumelanin and red pheomelanin. As we age, production of melanin drops, but black eumelanin drops the least. If your hair contains black eumelanin, your hair turns grey, whereas without it, your hair goes white.

What drove the evolution of dark and then light skin colour has been hotly debated and several competing theories have been proposed. Two of these are currently battling it out in the scientific literature for supremacy. The older of these ideas has been around for fifty years and revolves around the essential nutritional compounds folate and vitamin D. Folate, also known as folic acid or just vitamin B9, is an essential micronutrient that we need to repair our DNA and help cell division and a variety of other cell biological functions. A deficiency in folate gives rise to a range of medical complaints but at its most severe causes a variety of anaemia. It is particularly important during pregnancy and for the healthy growth of the foetus. A deficiency during pregnancy can result in premature birth and neural tube defects such as spina bifida, and for this reason there is strong evolutionary selective pressure to maintain folate levels. We know that when early prehumans evolved to live not in forests but on the plains of Africa, they lost their body hair and developed sweating in order to help maintain and regulate their body temperature. By losing our hair we exposed ourselves to more ultraviolet light which has been shown to cause folate to break down. As a response, the very early pre-humans evolved dark skin containing lots of melanin that absorbs more ultraviolet light and protects from folate depletion. This gets us to dark-skinned humans but the loss of skin pigment is then tracked to a different vitamin, notably vitamin D which is equally important for human growth. A deficiency in this prevents you from absorbing the mineral calcium from your food and leads to a number of bone-softening diseases like rickets in children and problems with the regulation of your immune system leading to auto-immune diseases and a susceptibility to infection. For most people in the world who don't eat a diet rich in fish, the main source of vitamin D is sunshine, and specifically, once again, ultraviolet light. But this time it is *con*structive rather than *de*structive and the ultraviolet light converts a version of cholesterol found in many foods into a precursor of vitamin D. The liver then does the final modifications to the chemical and makes vitamin D, which trundles off to the gut to ensure calcium is absorbed and to the immune system so that it can be properly regulated. So, when dark-skinned *Homo sapiens* first walks into Europe some 40,000 years ago, especially Northern Europe with its overcast skies and dark winters, the exposure to ultraviolet goes down and vitamin D production drops. Humans presumably struggled on suffering vitamin D deficiency until, at some point in the next 20,000–30,000 years, a mutation appeared that shut down melanin production. In Africa this would be bad news as the increase in ultraviolet destroys the essential folate, but in the gloomy north that is not a problem as there is so little ultraviolet anyhow. In fact, it is a benefit as now more ultraviolet gets through the skin and vitamin D production goes up.

This is the vitamin D-folate hypothesis for the evolution of both dark- and light-skinned humans. It is a lovely, neat hypothesis with straightforward nutritional demands that can clearly be seen as a driver for evolution. But some researchers believe there is no real evidence of this. A group based at the University of California in San Francisco have proposed that something else is going on. Previous studies had tried to recreate an ultraviolet-light-induced drop in folate in the blood when light-skinned volunteers, compared to dark-skinned ones, were exposed to ultraviolet light. The colour of skin made no difference and the ultraviolet light had no effect on blood folate levels. The conclusion was that ultraviolet does not penetrate the skin deep enough to have an effect on folate. Similarly, but in a different study, researchers looked at levels of vitamin D produced in another batch of volunteers as they went about their daily lives in Denmark in winter, when there is a very low level of ultraviolet. The vitamin D-folate hypothesis would suggest that light-skinned people could use this minimal ultraviolet light level and would make vitamin D, whereas the melanin in dark-skinned subjects would absorb all the ultraviolet and would have lower vitamin D levels. Yet the levels were the same in both groups. Admittedly, I am cherry-picking data here, as there is also evidence to counter this and support the vitamin D-folate hypothesis. However, the Californian group have concocted a different idea.

In their hypothesis, the big driver for the evolution of dark skin is water. Specifically, the pressures that built up when prehumans moved out of the forest, lost their hair and started sweating. Dehydration on the African savannah would have been an issue. Melanin in the epidermis makes skin less permeable to water and would therefore help conserve precious liquid. Then when dark-skinned humans moved into damp and dark Europe, while they still had plenty of vitamin D, they were now committing energy to producing melanin not needed for the conservation of water. Even though this may seem like a tiny expenditure of energy, similar features with unnecessary energy expenditures have been shown to drive evolution to select for the loss of the feature. Which could explain the loss of pigment from Europeans, for energy-saving reasons.

So, why did humans become dark-skinned and why did Europeans and East Asians lose much of the melanin in their skin? The only thing we can say for certain is that it's complicated. The folate part of the vitamin D-folate hypothesis is looking a bit shaky but many scientists have yet to be convinced of the water-saving benefits of melanin. The jury is clearly still out on this subject. We know it is something to do with how melanin absorbs ultraviolet light, but the exact mechanisms are yet to be decided. The very latest work on this is essentially trying to smush the two theories together and says it's a bit of both at the same time. Given biology is inevitably messy and intricate, it probably is something like this, but with even more complications thrown in.

The speed of death

In 2018 two scientists, Xianrui Cheng and James Ferrell from Stanford University in the USA, measured the speed of death at about 2 mm per hour (one sixteenth of an inch per hour). At least that is what the press release said they had done. It made for a very eye-catching headline and they received a lot of media publicity, which I'm sure made Stanford University very happy. Unfortunately, they had not measured what most people reading the headline may assume they had, namely the speed of death in a large-scale organism such as a person. What they had measured, though, was fundamental to all multicellular organisms on the planet, yet unknown to many people. It's a process known as programmed cell death, or apoptosis, an ancient Greek word meaning 'falling off', but, really, it's cell suicide.

> Apoptosis is not just found in the animal kingdom; plants have a similar mechanism. In the case of plants, though, the cell wall remains. Many of the structural elements inside a plant are made in this way. For example, the xylem tubes through which water moves inside the plant are made up of long-dead cells linked together and sculpted by apoptosis.

Now, on the surface it may seem peculiar that humans, and in fact all animals, have a process built into them that destroys healthy cells. Also, this is not some obscure rare cellular event. It is hard to estimate but inside you today, and every day, hundreds of billions of cells will undergo apoptosis. Why would the body have so many cells commit suicide? All cells within the human body have a limited life: red blood cells for example last for about 120 days, liver cells up to eighteen months, while white blood cells only hang about for thirteen days. There is a daily turnover of the 37 trillion cells in a human body of about half of 1 per cent. The problem that apoptosis solves is how to dispose of these defunct, worn-out cells. The inside of a cell is full of complicated enzymes and chemicals that could cause untold damage if allowed to wash around in the space between young and healthy cells. Having the cells that need to be replaced just explode and disintegrate would not be a good idea. The dismantling of cells has to be an ordered and controlled process.

Apoptosis was first described back in the nineteenth century, but it wasn't until the 1970s that we began to understand what was going on. Cells destined to commit suicide first receive a signal that is either generated internally or provided externally, usually from cells that make up the immune system. Once this signal is received the cell is doomed and a chain reaction takes place that starts chopping up the cell's DNA and initiating a peculiar process known as blebbing. The surface of the cell starts forming lumps, bumps and protrusions known as blebs. The tiny organs within the cell are pushed into these blebs that extend out, creating a weird star-shaped appearance. Special immune system cells that act like refuse collectors then engulf the blebs and safely break down and recycle the bits of the destroyed cell.

The science, and specifically the genetics, of how the death signal works was cracked at the end of the twentieth century by John Sulston, Robert Horvitz and Sydney Brenner. For their efforts they were awarded the 2002 Nobel Prize in medicine and yes, it really is called the death signal. The initial work was done in a tiny little roundworm species with the mouth-mangling name of *Caenorhabditis elegans*, or *C. elegans* for short. The reason they used this worm is that it is easy to keep, reproduces quickly and has a small enough number of cells that it is possible to count them all and follow their individual fates. What became apparent is that apoptosis was not just about the removal of defunct, worn-out cells. Cell suicide is regularly used as a form of cellular pruning. We now know that when the brain is developing in a human foetus, far more cells are produced than are needed in order to make sure that the right connections are made between brain cells. Cells in the wrong place or that have not made enough connections are then sent the death signal and carefully pruned out.

A lovely experiment was done at the turn of the millennium, in 2000, looking at how the toes of a foot develop in a frog tadpole. Careful observation showed that first a spade-like structure forms and this then splits into the individual toes. The clever bit of the experiment was to then stick a fluorescent tag onto any cell that had received the death signal. What became apparent was that the cells in the spade structure don't squidge themselves about to make toes. Instead, radial rows of cells commit suicide, leaving behind tissue in the shape of toes. Judicious use of apoptosis is an essential part of the development of the shape of every organ and structure in your body.

What is only just becoming apparent is how the death signal spreads around the cell and even between cells. Remember Cheng and Ferrell from the start of this chapter?

This is what they managed to see and measure: the speed of the death signal moving, in their case, across a large frog egg cell. Crucially, they saw that the death signal moved faster than simple diffusion would account for. Diffusion is the ordinary process by which a chemical will gradually spread out in a liquid through just random jiggling of the chemical molecules. If you put some salt at the bottom of a glass of water and leave it, in time the water will eventually become uniformly salty, without the aid of currents or stirring, just by diffusion of the salt. In the speed of death experiment, Cheng and Ferrell also had a red marker dye in the egg cell and could see that the death signal way outstripped the diffusion of this dye. It turns out that the death signal is spread using a form of positive feedback known to biologists as a trigger wave. This is a phenomena you see in quite a few biological systems, the transmission of an impulse down a nerve fibre being the best known and studied. While the exact mechanism for each example of a trigger wave is different, the underlying ideas are the same. An initial tiny signal, possibly involving a single molecule, induces a small change that itself produces more of the initiating signal. You now have more initiating signal, which produces more change and more initiating signal. This is a positive feedback loop and it causes the signal to spread much faster than diffusion alone. The current theory is that this system has evolved to ensure that when a cell gets the death signal, apoptosis takes place quickly and simultaneously across the whole cell to ensure the orderly removal of that cell.

However, apoptosis and cell death is not just of interest to developmental biologists as when the system starts to fail it can have dire consequences. The initial assumption about cancer was that it was a disease of uncontrolled cell division. A cell mutates, goes haywire, divides uncontrollably and turns into a tumour. However, now that we have a grasp on the genetics of apoptosis, and using the latest DNA sequencing techniques, it has become apparent that many cancers are partly if not wholly due to cells failing to respond to the death signal. Consequently, cells that should have been carefully removed hang about and produce a cancer. This opens up possibilities for tackling cancer. For example, some early experiments using capsaicin, the chemical that makes chillies hot, show that it can reactivate apoptosis mechanisms in prostate cancer cells in mice. Which is intriguing, but at the moment to get this effect in a person the daily recommended dose of chillies would be ten scorchingly hot habañera chillies eaten whole, including the seeds. This equates to 1,500 of the more popular but still hot jalapeño chillies. Clearly it is a work in progress, but reactivating apoptosis in cancer cells is an exciting possibility for treatment.

One final intriguing idea is worth mentioning about apoptosis. The genes involved in controlling apoptosis are usually very similar between related species, but not when it comes to humans. Compared to other primates such as chimps, our apoptosis genes are quite different; in particular, they are much less active. When this came to light it was suggested that it may be why we have such big brains. I have already mentioned that brain formation involves a lot of careful removal of excess cells by apoptosis. If, when compared to a chimp, human apoptosis is less active, this may explain why humans have such a large brain compared to our body size. Not only that, but it may also give a clue to the long lifespans of humans. While the speed of apoptosis in a cell may not directly relate to the death of the whole organism, apoptosis is certainly more than just a quirk of biology and may lead to an understanding of not only the evolution of our intelligence but also the process of our own death.

The difference between mostly dead and all dead

Most of us tend not to spend much time considering our own death. In many Western cultures a taboo keeps it out of polite conversation, so you might not have considered the tricky problem of what it means to be dead. On the surface this seems like a stupid thing to consider: after all, when something is dead it is dead and there is no coming back. While this is true, the problem is how and where you draw the line between death and life. When something is very alive or all dead it's obvious; the issue comes when circumstances bring somebody to the edge of that line of slightly alive or mostly dead.

In 1628, the English anatomist William Harvey described correctly for the first time how blood circulates around the body. His radical book took twenty years to be fully accepted, as it challenged scientific dogma that had persisted since the time of the Roman medic Galen. But in the book Harvey sets out the first biological concept of death. Until then, many medics were hesitant even to attend to the terminally ill, as their reputation, and thus livelihood, depended on living patients, not dead ones. Harvey set down the simple idea that when the heart stopped beating and blood stopped flowing around the body, the body was dead. It must be said that for such a simple concept Harvey spends an inordinate length of time explaining this in his book. This idea, known today as cardiovascular death, gives an attending medic or nurse an easy test to perform that can confirm the absence of life and the arrival of death. If a pulse is detectable then the person is alive, if not they are dead.

Except of course it isn't that simple. There are plenty of times in medicine when a very weak pulse will be undetectable, sometimes due to the person's physiology and sometimes due to an infectious disease. In the centuries after Harvey it became quite a problem, especially once a physician called William Hawes championed the attempted resuscitation of victims of drowning from the River Thames in London. In 1773, Hawes offered a reward to anyone who brought him an unconscious body taken from the river. He would then attempt to resuscitate the person and, it should be noted, the reward was paid irrespective of the outcome. His methods proved on occasion to be successful and people whose hearts had stopped beating, and were thus displaying cardiovascular death, were brought back to life. Clearly, cardiovascular death was not as final as Harvey and the medical institution had thought.

Through the nineteenth century, the question of when death has truly arrived was a major concern for many in the USA and Europe. It became ever more apparent that an absence of a detectable pulse was not enough to confirm death. This realization fuelled a proliferation of patents in the 1800s for safety coffins that all contained a variety of bells, breathing tubes and communication devices to ensure that you were not misdiagnosed as deceased and buried alive. It seems to have been a phobia that gripped the imagination of many, possibly explaining why the great American author Edgar Allen Poe dedicated five short stories to the subject including one, 'The Premature Burial', in which our protagonist creates an elaborate safety coffin for himself.

All this starts to change at the end of the nineteenth century. In 1887, Augustus Waller, a French-born physiologist working in London, recorded the world's first electrocardiogram, or ECG, that showed how the electrical activity of the heart changed over time. His first device was a far cry from the high-tech machines found in modern hospitals. It employed a clockwork toy train carrying a photographic plate that raced along a track past the projected image of a current flow measuring device. The resulting trace is unmistakably an ECG and the technology, once developed and the toy train replaced, opened the door for physicians to measure other electrical traces, such as those of the brain.

The first human electroencephalogram, or EEG, was recorded by a German psychiatrist called Hans Berger in 1924 and by the 1940s it had become a relatively routine procedure. So much so that it was possible for a physician to wheel out an EEG machine to check for brain activity at any time, for example to check for signs of life. The idea of brain death as a complement to cardiovascular death took hold and in 1968 France was the first nation to give brain death a legal definition. There is a conspiracy theory that swirls around the development of this idea that posits that the reason brain death was adopted by the medical community was that it helped source material for the newly perfected technique of organ transplantation. This theory suggests that medics were frustrated in their efforts to find suitable donor organs as once somebody was declared to have suffered cardiovascular death it was often too late to get the donated organ to where it was needed. Thus, a new definition of death was concocted that allowed patients to be kept alive after death until the organs were needed by the surgeons. While it is true that a person that has suffered brain death but not cardiovascular death can be kept alive by artificial ventilation and intravenous feeding, and the two technologies definitely arose at about the same time, a careful analysis of the history shows a different story. The correlation, as is so often the case in conspiracy theories, does not imply causation. The two fields of science advanced in parallel and only later come together to complement each other.

Having two definitions for death clears up most cases, but inevitably leaves a few edge cases still to be considered. Patients in a coma, or suffering from barbiturate or sedative overdoses, cannot exhibit any diagnostic reflex tests, such as the pupil reflex when a bright light is shone in their eyes, and can also give a negative EEG test. Which is why declaring somebody brain-dead while the rest of their body is still functioning is a tricky call for a medic.

What, then, is the actual cause of death for most of us? The short and not very helpful answer is a failure of homeostasis leading to cardiovascular death. The inside of your body and the body of any animal, a mammal in particular, has a very tightly controlled environment. The temperature of the core of your body is 37°C (98.6°F) with a normal variation of only half a degree up or down. Anything outside of this range is abnormal and often caused by disease or a failure of other bodily systems. It is not just temperature that is tightly controlled. The pressure, acidity and viscosity of your blood, the levels of oxygen and carbon dioxide inside you, and the amounts of minerals like copper, iron, potassium, sodium and calcium are also strictly regulated. Taken together, all of these control systems constitute homeostasis, which ensures that the chemistry that powers our bodies runs properly. All of the chemical mechanisms inside us have been optimized over millions of years of evolution to work in a very specific environment. Stray from that environment and the biochemistry either stops working altogether or becomes so inefficient that it may as well not be working. However, you will never see failure of homeostasis on a death certificate.

The Gaia Hypothesis, proposed by James Lovelock in 1972, takes homeostasis and applies it to the global ecosystem. Lovelock's hypothesis suggests that parts of our ecosystem interact to modulate the environment and make it suitable for life. Naturally, the influence of man can disrupt the Gaia hypothesis, just as external forces can disrupt a healthy organism.

The top two causes of death as would be listed on a death certificate in developed countries are currently heart disease and stroke. In both cases the underlying issue is the failure of blood flow to a specific region of our body, the heart and brain respectively. Consequently, there is a drop in levels of oxygen to those areas, normally carefully controlled by homeostasis, and the tissue dies leading directly to either cardiovascular or brain death. Within the developing world death from infectious diseases is incredibly high, the top suspects being malaria and conditions like cholera. With malaria, the infectious agent is a parasitic single-celled creature that hides inside blood cells, eventually destroying so many blood cells that the patient dies from anaemia, which in turn is essentially death by lack of oxygen. Cholera, on the other hand, causes terrible diarrhoea and is death by dehydration and an increase in blood viscosity. Even cancer ultimately disrupts the maintenance of the internal environment of the body by interfering with the correct functioning of an organ.

As Benjamin Franklin, or possibly Mark Twain, Daniel Defoe or Christopher Bullock commented, in this world nothing can be said to be certain, except death and taxes. While we may think that our own deaths will be of some unknown or unpredictable cause, the aphorism could be applied to the certainty of death by homeostatic failure.

If you need to focus then hold it in

On the evening of 8 December 2011, the leaders of the then twenty-seven European Union nations sat down for dinner in Brussels. The dinner was the start of a summit called to discuss the problems being felt by the financial institutions of the Union, especially those who had adopted the euro, the joint European currency created in 2002. As I sit writing this, the UK finds itself in the final paroxysms of

Brexit and its departure from the European Union. Looking back, the furore created by the December 2011 summit seems rather tame when compared to today. However, at the time it was big news. The UK prime minister David Cameron decided, foreshadowing what was to come, to invoke his ultimate negotiation weapon and deploy the British veto on the proposals being put forward. Cameron had gone into this meeting knowing full well that it was crucial for the UK and the politics of the time. So, to ensure that he was at the top of his game he deployed his own personal tactic for success. He refrained from micturition, which is the unnecessarily obscure scientific term for urination. To put it bluntly, he didn't go for a pee, despite needing one. His theory was that by holding it in and forcing himself to endure the discomfort of needing to pee, he focused his mind and made himself a more effective negotiator.

It was a technique that he had deployed before, most notably at the Conservative Party conference in 2007. In this instance he knew that he needed to deliver a stellar performance in order to rally the flagging party and boost it in the eyes of the voting public. According to subsequent documentaries made about the event, he took the possibly risky decision to memorize the carefully prepared speech and deliver it without an autocue. If he could pull off this tactic, it would add a dynamism to the performance and bolster the impact. He further decided to use the micturition avoidance tactic and held it in for the duration. In the end, history tells us that it was a big success, the speech was well received and the Conservative party benefitted from a much-needed boost to its ratings. Which is why he repeated the tactic for the European summit. It was not, however, something that Cameron had come across on his own.

It took some persistent investigative journalism to ferret out that Cameron was introduced to the technique ten years previously. One of the greatest, if deeply controversial, Conservative orators was Enoch Powell, infamous now for his blatantly racist 'Rivers of Blood' speech. When asked about his oratory tactics he explained 'you should do nothing to decrease the tension before making a big speech, if anything you should seek to increase it'. In Powell's case this included micturition avoidance and performing with a full bladder. It's clear that not going for a pee is a well-established technique for honing your ability to deliver speeches, but is there any scientific justification for this?

It turns out that a small amount of work has been done on this strange phenomenon and scientists who performed the early experiments have even won a Nobel prize for their efforts, although admittedly it was an Ig Nobel prize. The Ig Nobel prizes are awarded annually by the Annals of Improbable Research for scientific work that makes you laugh and then think. They started in 1991 and this year alone includes awards for studies on using roller coasters to hasten the passage of kidney stones, the efficacy of saliva as a cleaning agent, and using voodoo dolls as a means to retaliate against abusive bosses. In 2011 the Ig Nobel medicine prize was jointly awarded to two groups, one from the Netherlands and the other from Melbourne in Australia. In both cases, the groups looked at the effects of holding off going to the toilet on a variety of tasks that require concentration. The known biology behind this is that as your kidneys do their work of removing waste urea, unwanted salts and excess fluids, urine is produced that is stored in the bladder. When the volume in the bladder reaches about 400 ml (14 fl oz) the sensation that you need to go for a pee begins. Fortunately for us all, once you get to a few years old you develop the ability to resist the need to empty your bladder and can choose when and where to do this. The human bladder can easily hold twice this volume of urine and on waking in the morning may contain nearly a full litre of liquid (34 fl oz). Once the bladder has reached the critical point of 400 ml we experience mild sensations that you could describe as almost a feeling of being internally tickled. If you choose to ignore this sensation it gradually builds as the bladder expands to accommodate more fluid and its stretch receptors detecting this begin to report pain to your brain. Continue avoiding the need to pee and the pain increases until it becomes debilitating and ultimately accidental release occurs and you lose control over your bladder. Thankfully, most of us rarely encounter this situation and it is wise not to put yourself in this circumstance too often, as repeated and prolonged holding of urine can lead to urinary tract infections, as bacteria are given time to multiply and cause problems.

Of the two Ig Nobel prize-winning groups, the Australians tested eight individuals by having them drink 250 ml (8 fl oz) glasses of water every quarter of an hour until they reported either a need to urinate, a strong need or an urgent need. At each point, they then tested the subjects using a cognitive function test to see how the urge to pee was interfering. The results showed no change in the ability of the subjects to perform the tasks until they hit the urgent need to pee level of pain. At which point the performance drastically reduced and once they had relieved themselves the performance predictably bounced back up. Contrary to this negative finding, Mirjam Tuk from the Netherlands showed that for some tasks needing to pee improved performance. In this study participates were told they were performing a water tasting test. Half the group were given a full 700 ml (24 fl oz) of water to drink in several cups and the others told to take just a few sips. They were then forced to endure a forty-minute wait, the time it takes for water to be absorbed and excess fluid to be processed by the kidneys, before being given some choices. A financial award was offered for participating in the experiment, either €20 to be handed over the next day, or twice that amount in a month's time. She then asked the now presumably somewhat confused experimental participants how badly they needed to go to the toilet. The people who had initially drunk the large volume of water reported on average a need to pee of 4.5 on a scale of one to seven. Those that had taken sips had no significant need to pee. This need to pee rating correlated to who took the money the next day and who was willing to hold out for the bigger financial reward. Those with a full bladder were more likely to take the larger amount of money in a month's time. It would seem that in this experiment at least, having a full bladder increased your ability to resist quick gratification and hold out for a bigger reward. These results were a bit surprising, as they go against the idea of ego depletion that should have held sway in this case. According to this theory, each of us has a finite pool of self-control at any one time. If you use this up then you will be less likely to resist another call on your self-control. In Tuk's experiment, by resisting the need to urinate the test subjects should have been more likely to take the sooner rather than the later financial reward. To explain the findings, Tuk has subsequently developed a theory of inhibitory spillover, which reasons that because the call on the subjects' self-control was simultaneous, the ability to hold out for more money was aided by the already established pattern of resisting the need to pee. While this effect may seem at first glance to contradict the joint Ig Nobel winners from Australia, the two are complementary. The Australian work only showed a negative effect when the poor test subjects were suffering an extremely urgent desire to urinate. The effect of holding off going to the toilet seems to be something that builds, increasing your self-control, and then crashes once the need for the toilet gets too great.

While this may seem like a tenuous result, further studies have confirmed the inhibitory spillover effect. A group in California State University set up an experiment to see how well people could lie when they also had a full bladder. Some twenty-two students were given questionnaires to establish their true beliefs on a number of controversial social and moral issues. They were then interviewed by a panel and instructed to lie about their feelings and beliefs on two of these subjects. Half the students were given 700 ml (24 fl oz) to drink forty-five minutes before the interview and half just 50 ml (1¾ fl oz). The group that drank the larger amount of water all had full bladders and a desire to urinate by the time they were interviewed. When the results came in, the water-drinking and desperate-for-the-loo students were more convincing liars who gave longer, more fluent and crucially more believable responses to questions they had been told to lie about.

Combining this work with the two Ig Nobel winners we could infer that the inhibitory spillover helps not only your ability to exhibit more self-control but also an increase in ability to process complex ideas. Lying well is a really taxing task that puts a large cognitive load on our brains and it is possible that needing to pee helps by focusing your mind (see page 153 for more on lying). Tuk went on to look at how ego depletion and inhibitory spillover worked together when she moved to Imperial College in London. The latest experiment asked people to watch a film and then answer some questions in a short interview. The subjects presumed that the focus of the study was the film, but unbeknown to them what was really important was the bowl of potato crisps sat on the table during both parts of the test. Half the subjects were asked to exert control over their emotions during the film and half were just shown the film without instruction. The subjects exerting emotional control showed inhibitory spillover and resisted eating the crisps during the film but then ate more during the interview. Sadly, Tuk didn't try getting the volunteers to drink first so we don't know if a full bladder would help you resist the crisps.

So, when David Cameron decided to apply the advice of Enoch Powell and keep a full bladder while he was conducting those crucial financial negotiations with the European Union leaders in 2011, there was a sound scientific basis for this. By forcing himself to apply self-control and not dash to the lavatory, it seems that the inhibitory spillover effect would come into play. His ability to deal with complicated interactions with other people would be enhanced and it presumably gave him an edge in his performance. While this effect would also, according to the research, make him a better liar, it would not, I'm sure, be of any significance. Anecdotally, I can add my own experiences and I am happy to admit that I get the most writing done in the late afternoon, usually after constant cups of tea, when I am trying to reach a suitable level of achievement for the day. Significantly, by this time I also invariably have a considerably full bladder. The secret to this technique would presumably be to catch the benefits of the inhibitory spillover, enhancing your ability to focus, but avoid the negatives discovered by the Australian work and not let yourself get desperate. The lesson we learn is that you should not leave it too long before you find relief from the need to pee.

BEING VIRTUALLY HUMAN

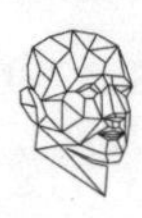

The uncanny valley and being nearly human

At the end of the 2016 Star Wars franchise film *Rogue One*, and this isn't really a spoiler, the actress Carrie Fisher makes an appearance in her role as Princess Leia. But it's not Carrie Fisher. For reasons that are too complicated to explain, and would involve spoilers if you haven't seen it, what you see on screen is the version of Princess Leia as depicted in the original *Star Wars* film made way back in 1977. Using the latest digital technology, the face of Leia was lifted from archive photos and animated. When I saw the film, it had a peculiar effect on me. Firstly, I was quite upset because Carrie Fisher had sadly passed away the very day before I was sat in the cinema. But the other effect it had on me was that I didn't believe what I was seeing.

I knew it wasn't really Carrie Fisher on screen but a computer-generated image. And it bothered me. It didn't look right and was a bit unsettling. I didn't have that feeling in the rest of the film, despite all the crazy alien species and starships blowing up. The animated young Carrie Fisher as Princess Leia had fallen into what was first called *bukimi no tani genshō*, which translates from the Japanese as 'eerie valley phenomenon', and is now called the 'uncanny valley'.

This idea was first laid out by Professor Masahiro Mori in 1970 and the Anglicized term 'uncanny valley' was first used in 1978. The theory is based on a few observations, starting with our ability to empathize not only with other humans but with pictures of humans. What's more, as an image of a human being becomes ever more realistic we find it increasingly easier to empathize with. This is particularly true for faces. If you start out with a simple smiley face emoji – two dots for eyes, one for the nose and a curve for a smile – it doesn't really convey emotion. My phone now allows me to use a smiley emoji that is shaded, has coloured skin, teeth and even eyeballs with pupils and eyebrows. This conveys more emotion, and this effect continues as the face becomes more lifelike. Now, you may predict that this keeps going and as the face gets closer and closer to being a real photograph of a human, my perception is more and more empathetic. Except what Professor Mori realized is that when you get really close to reality, but it's not quite perfect, that feeling of empathy suddenly drops away and is replaced by a sense of oddness, or even revulsion.

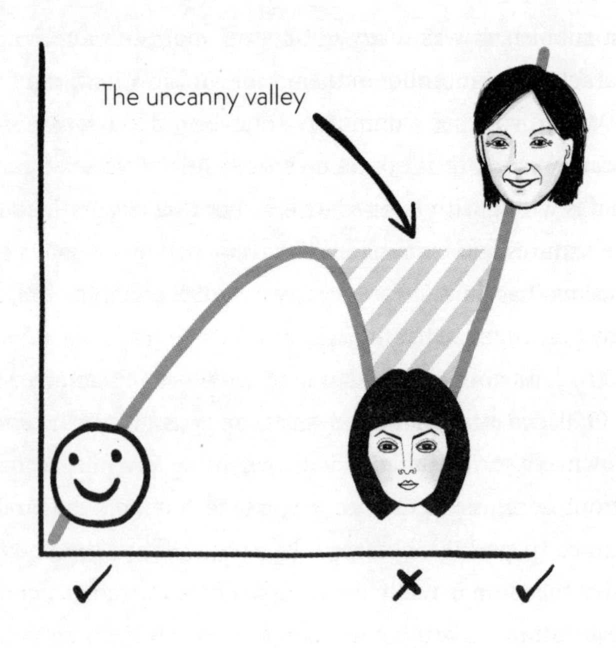

If you return to some of the earlier animated films that attempted to make their characters as real as possible, you can see this problem very clearly. In 2004 a children's film called *The Polar Express* was released. This featured the voice of the American actor Tom Hanks, was completely computer animated and tried to emulate reality as close as possible. The characters were not cartoons, and the audience was supposed to see them as real people. It did not get great reviews from the critics. A typical reviewer described it as 'at best disconcerting, and at worst, a wee bit horrifying'. The computer animation was cutting edge and the best that they could make it, so it wasn't as if the animation was just rubbish. It was really good – so good, in fact, that the characters fell into the uncanny valley.

When you see computer animation that falls in the uncanny valley it is sometimes hard to put your finger on what is wrong with it. There will be nothing obvious about it. The textures and expressions are all meticulously copied, but it seems that humans are very, very good at spotting a fake. Why this might be has given rise to some interesting theories. After all, it's not as if humans evolved with this being an issue.

One idea is that our brains are getting tangled up in what is known as a sorites paradox. I'll cover this more fully when we get onto discussing the peculiarities of human behaviour in a crowd (page 178), but from the perspective of the uncanny valley the issue is really that our brains are trying to answer the question, 'When is a human face not a human face?' This explanation of the uncanny valley suggests that when your brain sees something that it can just about detect as not human, it gets confused. We are very good at detecting faces, it is hard-wired into us, and babies from birth can recognize a human face. We can tell when a face is a bit off and this sets up a cognitive dissonance inside our heads. Our brain is trying to put what it sees into a box – is it a face, or is it a pretend face? – but it can't make up its mind and we are left with an eerie feeling of it not belonging in either box. An obviously cartoon face or a real photo don't give us this issue, as we can easily decide which box to put it in.

An alternative idea tries to explain why we feel revulsion at seeing an uncanny valley animation. Humans are not just hard-wired to recognize faces, we also have a built-in antipathy to things that may cause disease. This is why rancid meat or rotten food is inherently disgusting. We see it and a deep-seated feeling is aroused that makes us want to avoid it. From an evolutionary perspective, this is beneficial, as we then avoid things that could make us ill. The idea has been suggested that when we see a face that falls within the uncanny valley, we detect that something is not right, and it sets off the alarm bells. We perceive it unconsciously as being diseased, ill or potentially harmful and that manifests as revulsion.

The basic ideas behind the uncanny valley can be traced much further back than 1970 and Professor Mori. In Charles Darwin's book, *The Voyage of the Beagle* (1839), he describes a South American pit viper as being peculiarly ugly because the positioning of its facial features mirror those of a human and thus give it a 'scale of hideousness'.

The idea of the uncanny valley has some serious implications for future technology. Will we ever be able to make truly realistic computer animations, or even robots, that don't fall in the valley? This is not just a question for media producers and game players. Computer animation, especially computer-generated 3D virtual animation, is finding new applications in areas like medicine. Fully immersive virtual reality simulations are being used to treat post-traumatic stress disorder, helping paraplegics regain muscle control and even as part of pain relief for burns patients. The worry is that if animations fall into the uncanny valley they may hinder rather than help the patients. One of the leading researchers in the field, Dr Angela Tinwell from the University of Bolton, has suggested that we may never climb out of the valley and it may be an unscalable uncanny wall. In which case our only solution will be to back off and come out of the other side of the valley where we can then create robots and animations that are clearly not human, but at least they won't be revolting.

How language makes the world go round

What makes human beings what we are and what makes us so different to the rest of the animal kingdom of organisms? On some basic biological levels, the answer to this is nothing. We are all just like any other organism, a collection of atoms arranged into molecules, gathered into cells and organized as the product of blind evolution. And yet there are some things that *Homo sapiens* does better and differently than any other species. Our ability to communicate is second to none. Now, you may be thinking, 'What about dolphins and the way they speak to each other through whistles and clicks?', or 'How about a hive of bees that coordinate their efforts through complex bum-wiggling dances and chemical messages from the queen bee?' But even if we are generous with the extent of the ability to which other animals communicate, human beings are still, by far, the masters of the art.

Key to our ability to convey messages from one to another is clearly the spoken word and the evolution of our linguistic abilities. There are many opposing views on exactly how *Homo sapiens* learned to speak, but they can be categorized into two broad camps. One group of theories suggest that our early hominid ancestors gradually developed sequentially more complex communication strategies. Presumably we started with grunts and gestures that gradually took on specific meaning and sophistication, until different grunts became words and a grammar arose that allowed us to place those words into a larger context and convey ever more complex ideas. These continuity-based theories are by far the most popular among linguists such as the renowned Canadian-American psychologist Steven Pinker. On the other side of the evolution of language debate are a few discontinuity-based theorists, with Noam Chomsky, the eminent US linguist, as their champion. Since the 1960s Chomsky has held that as our abilities with language and grammar appear to be instinctive in young children and babies, this implies that language is genetically hard-wired into us. If this is the case, there is an argument that it arose in a relatively short time period, in large jumps rather than in a slow, gradual and continuous process. Either way, human language is at the heart of our ability to produce culture and transmit ideas to one another, both of which drove our technological development which is now, it would seem, developing its own languages.

In November 2016 a group of computer scientists based at Google in Silicon Valley announced that something strange was going on with the new artificial intelligence system they had recently installed to run their Google Translate online service. Previously Google Translate had used a system known as example-based machine translation or EBMT. This older method used a huge library of texts that had been translated into two different languages, with the complementary words and phrases noted in these translations. When the algorithms were then presented with a new, unknown text to translate, they used Google's well-established and blindingly fast search routines to sift the library of already translated texts to find analogous chunks. Once these chunks were identified, the matching translated chunks could be extracted and assembled into a full translation of the unknown text. It is a somewhat crude way of translation that does not rely on any underlying understanding of the languages being translated, yet it produces pretty good results.

In September 2016 Google had switched a couple of its translation services to a new way of thinking. Neural machine translation, NMT or what most people would call artificial intelligence, was developed and set loose on the English / Korean and English / Japanese systems. The artificial intelligence was trained by feeding it thousands upon thousands of translated texts until it was able to competently turn Japanese into English and back again, and Korean into English and back. So far, so good, and nothing particularly innovative there. The older example-based machine translation system could already do this for 103 different languages to and from English. However, if you needed to translate from French into German, the older system would translate first from French into English and then from English into German. It always uses English as the lingua franca or common language. Which is where the odd thing happened with the newer Google translation artificial intelligence. Once the computer scientists had the neural machine translation up and running with English / Korean and English / Japanese, they set it an altogether more challenging task of translating Korean to Japanese directly, without going first through English. The artificial intelligence managed to produce a reasonable translation, not as good as the example-based system, but what you need to remember is that the new system had never been taught to do this translation. It had worked out how to do it without human intervention.

Clearly the Google researchers were delighted that their new artificial intelligence was exceeding its programming, but discovering how it had done this, or at least how they think it had done it, came as even more of a surprise. The problem with artificial intelligence systems is that once they have been running for a while the programmers don't really know what is going on inside any more. The neural machine translation had evolved its own programming and left its creators behind. However, they do have ways of peering into the computer code and based on this analysis they determined that the neural network was creating its own associations between similar concepts in different languages. Similar ideas were being grouped and given their own classifications unique to the computer. The researchers described this finding as the computer system having created its own 'interlingua'. The understanding is that the artificial intelligence is using its interlingua to make the leap from Korean directly to Japanese, without going through English as all previous algorithms had done by nature of their fixed programming.

> There are between 7,000 and 8,000 different languages currently spoken. This seems like a big variation, but it turns out to be very difficult to nail down the difference between a distinct language and a dialect. Of these languages, about 5 per cent, that's just 380 languages, account for 95 per cent of the primary tongue of the world's population.

When this was announced some commentators reported with significant fanfare that the Google algorithm had created its own new language, a private language that only it spoke. Others were less sure and dispute that what the artificial network had done was language creation but rather just grouping of meanings. Interestingly, this in itself is a semantic argument, as while the new computer interlingua does not display many of the traits of language – grammar and syntax, for example – it is performing the same function of a language, though admittedly that is being used by an intelligence that is somewhat alien to us. In some ways, what the interlingua is doing is more fundamental to language than what we presume occurs in our brains where we have a concept and associate it with a word. For those of us lucky enough to be fluent in more than one language, something like this must be happening in our own brains.

The Google translation interlingua is not the only example of computers messing about with language, as another of the tech giants has been looking at this too. Computer-science researchers working with Facebook have been looking to develop an artificial intelligence that human users could interact and talk with, initially through text, in a way that seems convincingly human. A simple and convenient way to do this is to create two artificial intelligences and allow them to talk to each other and thus learn from their own interactions, running at a speed that no human could ever match. But, as with any computer program, you have to be very careful to specify exactly what you want as the end result. After trying their system a few times, the programmers realized that they had failed to stipulate a very important criteria for success. The two system that were set up were named Alice and Bob and were set the task of negotiating between each other who took possession of a number of virtual items: hats, books and balls. Alice and Bob each had their own agenda and placed different values on the different items. While the computers began chatting to each other happily using English, trading items back and forth, the programmers had not ruled that they should stick to English. Soon the artificial intelligences were producing what looked like complete gibberish. A typical exchange about who had the collection of virtual balls looked like this:

> Alice: balls have zero to me to me to me to me to me to me
> Bob: you I I I I I everything else
> Alice: balls have a ball to me to me to me to me
> Bob: I can I I everything else

Despite this nonsense being spouted by the computers the end result of these bewildering negotiations was a successful distribution of hats, books and balls. Since the programmers had failed to restrict Alice and Bob to using English they had developed their own shorthand ways of communicating with each other. In the above example we think that the repeated 'to me to me to me' and the 'I I I' indicate the number of items Alice and Bob are offering each other. I say that we think this is the case because just as with the Google interlingua we are not completely sure, as this peculiar dialect of English is only spoken by the computer Alice and computer Bob and there are no translation rules for it. While not as sophisticated in its scope as Google's neural machine translation, Alice and Bob were on their way to creating their own language – until they were turned off. The aim of the Facebook research was not to create a private language for two computers but a system that could interact with human beings in a human language. So, the experiment was stopped and Alice and Bob were turned off at the mains and the nascent language died.

The ability to communicate is fundamental to what it means to be human. It has allowed the spread of ideas, culture and technology between individuals which has in turn allowed us to be as spectacularly successful as we have been. With the advent of computers and then the worldwide network of computers we have been able to communicate, for good or ill, faster and with more people. If we can give computers the same ability it could potentially open up new vistas of social, cultural and technological development for us. But how comfortable are you if you cannot understand the language computers use? There is a risk to the development of artificial intelligences as has been expressed by influential people like Bill Gates, founder of Microsoft, and physicist Stephen Hawking. In 2014 Hawking summed up the fear when he said that 'developing artificial intelligence would be the biggest event in human history. Unfortunately, it might also be the last.' If we go down this route we may want to keep the neural networks talking in a language at least some of us can understand.

Behaving virtually bad

In 2017 a survey of 4,248 adults in the USA found that some 41% of them had experienced some form of online harassment. Dig deeper into this data and you find that half of this behaviour falls into the category of serious intimidation: threats of violence, stalking and sexual harassment. To put that in context, that means in every group of five people, one will have experience of low-level online abuse and one will have suffered more serious abuse. Except, of course, it is not that simple and if the group of five people are white men then probably none of them will have any abuse experience. On the other hand, if the group is of black or Asian women then there is a good chance they all have a tale to tell about online abuse targeted at them. These figures may not come as a surprise to you. If you delve below the description line on sites such as YouTube and take a look at the comments, you quickly realize that it seems there are a lot of really horrible people online who are willing to say vile things. And yet you don't experience this same antisocial behaviour to this extent in your day-to-day offline world. The vast majority of people are perfectly pleasant to each other and despite potential disagreements don't say or behave in such hateful ways. So why are we so horrible to each other when we are online? This phenomenon has long since been identified and is called the *online disinhibition effect*. Psychologists can break down the cause and effect of why and how this happens, but the bottom line is that when we are online we have less reason to be cooperative and be nice to each other. The inhibitory effect of millennia of evolution that keeps us civil is removed and we turn into jerks.

A 'troll' is a common term for those who indulge in behaving badly on the internet. It originates in the early 1990s on a pre-World Wide Web bulletin board called Usenet, where people habitually went 'trolling for newbs'. By posting a provocative and repeatedly discussed question on the bulletin board, regular users could identify the newbs.

You may think that this is an overly optimistic view but there does appear to be an underlying human trait towards cooperation and working together. Researchers at the Human Cooperation Laboratory at Yale University in the USA designed a series of experiments to test this idea. Groups of four participants were brought together online and each given some money, a single US dollar, for example. They were then asked to make a choice to either contribute the money to a central pool or keep the money for themselves. Since the game is played online, nobody knows what your choice is and your anonymity is assured. After each choice any money in the central pool is doubled by the generous game administrator and then equally shared between all the players of the game. The only other rule is that if nobody puts any money in the central pool, you all lose your money and get nothing. So, if everyone contributes their dollar the central pool contains four dollars, which is doubled to eight, and each player is given two dollars. So, you all lose a dollar to the pool but all gain two dollars and everyone wins. But if only one person puts a dollar into the pool, and the other players selfishly hold onto their money, that one dollar is doubled to two and each player gets a half dollar. If you were the sole person to put the dollar in you end up with just half a dollar, while everyone else has one and a half dollars. If two people put money in the pool they will end up with a dollar and the selfish players get two, while if only one person is selfish the people who put money in the pool get one and a half dollars while the selfish person gets two and a half. The game is a kind of four-way prisoners dilemma, where if you want to make the most money you need to be the only selfish person in the group but the second best option is to be fully cooperative. What makes the game interesting is how the results change if you give the players time to think about their choice. When the Yale scientists gave the players time to think about what they were going to do there was a greater chance that they would not cooperate, but when they were made to make a decision within just a couple of seconds they usually went for cooperation. This has been shown to be the case with cultures around the globe in different experiments – the gut reaction of human beings is to play nicely.

This makes perfect sense when you consider our evolutionary and developmental past (see page 3). Our species *Homo sapiens* evolved in small groups of individuals for whom working together in a harsh and unforgiving environment was crucial for survival. Rather than lingering over possible marginal gains from being selfish, we were best served by helping each other and it is believed we have evolved to be essentially generous to one another when making gut choices.

On the surface, this makes the behaviour of people on the internet all the more bizarre. Something is removing us from the built-in desire to work together and cooperate. It is not rocket science to work out that it is the anonymity of the internet that allows this to happen, except it is more than just being anonymous. A group of Israeli scientists tested how pairs of people interacted when presented with a thorny moral question to discuss online through a chat messaging system. The volunteers were split into four categories: those that were just interacting through a text chat system; those who revealed their names, but nothing more; those who could see each other on a webcam; and those that could actually make eye contact on a closer webcam. The biggest impact on maintaining civility in the discussions was when the subjects had eye contact. If somebody knows your name that is not enough to prevent uncivil behaviour, something that was borne out in South Korea in 2007 when a law was enacted designed to prevent anonymous use of social media platforms. Despite people being no longer anonymous, in so much as it was possible to find out who an online account belonged to, it only reduced online abuse by less than 1 per cent. Given these marginal gains in abuse reduction and the significant burden it created for online sites the law was subsequently overturned.

This poses a problem for tech companies and governments trying to stem the flooding tide of internet abuse. If we humans are evolved to cooperate with each other but only when we can see each other, how do you civilize an online system that is inherently removed from the presence of other people?

The answer may lie in understanding another aspect of online culture, namely its ability to create groups of shared belief. Given the vast reach of the internet and our ability to search through it with lightning speed, it has become possible for people with niche interests or beliefs to find others who share their passions. This in itself is a mixed blessing, as for every support group for parents with a child suffering a terrible and rare disease there is an extreme religious or political group intent on causing real-world harm. The internet has allowed these groups to form, as finding like minds has become trivial. In a group of people that meets face-to-face we form bonds within the group through shared actions, clothing, expression, jokes and mannerisms, but online all we have is the written word. Consequently, what is written must form the function of all of the other things that would normally glue a group together. In some cases, online abuse appears to be less about specific targets and rather more to do with signalling that you are part of a particular group. If you add to this that online posts that either trigger or express outrage are more likely to be forwarded on, you set up a positive feedback loop of behaving badly. In order to seem part of a group and receive validation from your peers you are best served writing something that gets passed around, such as an outrageous and unpleasant slur. Saying something conciliatory, uncontroversial or just unexciting will get you nowhere. The online vileness is just a way of having yourself seen as being in the club, which opens a way to influence people to be less vile.

If part of online abuse is to do with the social dynamics within a perceived online group of people, then given its anonymous nature it is possible to turn the system on itself to encourage less abusive behaviour. Kevin Munger, while working on his PhD at New York University, set about coding a series of automated Twitter robots, or bots. First, he identified the people that these Twitter bots would be targeting by looking for those who regularly tweeted abusive racist comments. He then set up his Twitter bots with fake followers to give them the air of legitimacy and so that the bots would appear to be of a higher status within the targets in-group. Then he waited for the targets to tweet something abusive, at which point the bots jumped in and replied with a gentle rebuke phrased as if coming from a believable real person. While initially this resulted in a stream of abuse targeting the bots, the effect of being admonished by those they saw as higher up in the Twitter pecking order resulted in the targets moderating their language for some considerable time.

For some people the internet, and in particular social media, has become a toxic swamp of abuse. If for whatever reason you fall under the glare of an active group of online abusers you could be on the receiving end of a deluge of horrendous insults. It would seem that the internet has allowed us to sidestep the evolutionary imperative for human beings to behave nicely to each other. The result being that with this safety brake removed, some of us begin to behave like jerks. What remains to be seen is how we can bring the behaviour back in line with face-to-face interactions and make the internet and social media a pleasant place for us all.

Getting your internet fix

Dr Ivan Goldberg was a psychiatrist living and working in New York in the USA. In 1986 he decided to use the newfangled internet to set up a bulletin board called PsyCom.net that would allow his fellow psychiatrist colleagues and himself to discuss matters of interest from the comfort of being in front of their own computers in their own offices. The online community was a success and grew. A topic of regular discussion was the American Psychiatric Association's book called *Diagnostic and Statistical Manual of Mental Disorders*. This tome gives the psychiatric profession in the USA a series of benchmarks against which to measure any behaviour the practitioners observed. In 1995 Goldberg posted a short parody of the dusty and dry prose to be found in the manual every US psychiatrist had on their shelf. The topic of the parody was an entirely invented disorder he concocted called IAD, or internet addiction disorder. To his horror, rather than chuckling at his post, several of his colleagues replied to say they feared they may well be suffering from IAD themselves. He received emails asking for help and soon hundreds of self-diagnosed addicts posted on the bulletin board to share their problems. Goldberg's invented disorder turned out to be only too real.

Or is it? In over twenty years since Goldberg invented internet addiction the consensus on whether it is a real thing or not has swung back and forth. At the moment, though, there is mounting evidence that internet addiction is a bona fide condition and a real addiction. It is now seen as part of a larger group of behavioural addictions that include gambling, porn and online gaming addiction.

Much of the pioneering work on internet addiction is coming out of South Korea, one of the most internet-connected countries on the planet: more than 80% of Koreans have a smartphone; 92% of the population uses the internet on a regular basis; the country has the fastest average internet speeds in the world; they have the best peak speeds available to home users; and they were the first to launch a super-fast 5G mobile network. In South Korea, the internet is a pervasive and ubiquitous tool available to all at speeds many of us can only dream about. However, this technological prowess does not come without its disadvantages.

The Internet and Communications Technology Development Index is an annual United Nations report, rating countries on three criteria: access to the internet, how much the internet is used, and how good people are at using it. For several years the top-ranked country was South Korea, but in the latest report it fell 1.5 per cent behind the new leader, Iceland.

A survey carried out in 2013 found that 7 per cent of South Koreans between the age of five and fifty-five were at risk of internet addiction. Given the population of over fifty million, that makes for some three and a half million possible addicts. If you narrow the age range the data is even more alarming. Among teenagers the figure is at its highest, with one in eight being at risk of internet addiction. It comes as no surprise that South Koreans are working hard at identifying what is going on.

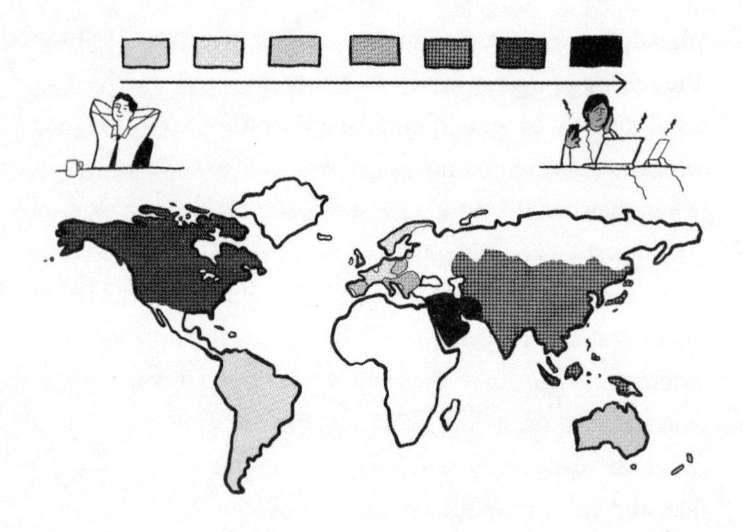

Global internet addiction

Addiction was traditionally looked at as something that involved the abuse of a mind-altering drug such as cocaine, heroin or alcohol. The use of the drug sets up a dependency whereby long-term changes in the chemistry of the brain not only alter how the drugs affect the body, usually by diminishing its effect, but also by making it hard to stop using the drug as the withdrawal symptoms are so unpleasant and possibly dangerous. It took many years to really reach an understanding of how addiction works. An early experiment from the 1990s was for a long time ignored. Kent Berridge, working at the University of Michigan in the USA, fed rats a sugary syrup and noted that when he did so they would lick their lips in pleasure and keep coming back for more. We already knew that the sensation of pleasure was controlled by the release of dopamine inside the brain. But when Berridge used surgery to knock out the dopamine-producing parts of the brain, the rats no longer wanted the sugar syrup, but when they had the syrup they still licked their lips in pleasure. The rats still liked the sugar but no longer wanted it. This was a key discovery for the science of addiction as it dispelled the myth that addicts both liked and wanted the source of their addiction. While the two emotions of like and want are usually correlated in us, it does not need to be so. A drug addict will want the drug even when they have long since ceased to like and enjoy the act or results of taking it. Similarly, with a behavioural internet addiction the user may no longer enjoy their time online but they keep doing it as there is a more powerful want to carry on. The release of dopamine is more to do with our want to have something than to do with our liking the act of having it. If an activity creates a dopamine spike within your brain, for whatever reason, be it an activity or a drug, there is the possibility that it could lead to addictive behaviour.

Careful examination of the things we do on the internet has given researchers an idea as to what makes a particular activity potentially addictive and likely to give the dopamine spike, at least when it comes to computer and online games. Take, for example, the mobile-phone game called *Flappy Bird*, designed by Dong Nguyen from Vietnam. According to the creator, the game took a couple of days to make and was supposed to be played as a little time filler for when you were relaxing. The game was released on Android and Apple devices in May 2013 for free, but included adverts that would in theory make the creator money if it became popular. For the whole of 2013 the game did not prove to be a success and made no money but then at the start of January 2014 downloads of the game rocketed and it became the number-one title on both mobile platforms. According to the maker, at its height of popularity the game was bringing in US $50,000 a day. But then just a few weeks into its success, Nguyen announced that he was withdrawing the game. He had realized that his little time-filler game was an addictive monster. The game is ludicrously simple: you control a flapping bird moving left to right on the screen – when you tap the mobile phone's screen, the bird flaps and is pushed upwards but then immediately begins to lose altitude and fall back down. You need to constantly keep tapping to keep flapping and keep the bird aloft. The aim is to avoid the peculiar green pipes protruding up and down on the screen and navigate your bird through the gaps. And that's it: you score a point for each gap you negotiate, the pipes to dodge just keep coming and the challenge is how high you can get your score. It is really hard and, on the surface, does not look like it would be that popular, but it is, and I just lost an hour playing the game in the name of research so that you don't have to. Despite being very bad at the game, crashing my flappy bird almost immediately, I kept restarting for just one more attempt. What makes the game successful is that it hits upon many of the things identified as crucial by behavioural addiction studies. Firstly, it provides simple goals that always seem just out of reach. Each gap that you navigate the flapping bird through is no harder than the one before, the issue is just how long can you keep going. Secondly, you get better at the game as you go on and see progress in your score. The more you play the better you get and the higher your score and sense of achievement. Finally, there is feedback, which seems to be crucial for the development of behavioural addiction. We have known about the game-changing nature of feedback for a long time, since 1971 when Michael Zeiler from Emory University, Georgia, USA, published some work on pigeon pecking. It was a simple experiment using just three white pigeons. An automated feeding system was set up in the birds' cages that was linked to a button. When the pigeon pecked the button, they received a reward of delicious pigeon pellets. Every now and then the pigeon would wander over, peck and a pigeon pellet was delivered. Zeiler then added a random element into the experiment. If the button only delivered the food on average once every ten pecks, the birds gave up and ignored the button. But if the reward was given seven times out of ten pecks, the pigeons became obsessed with pecking the button and would spend twice as much time pecking and eating. Zeiler had produced a behavioural addiction by simply adding in a random chance of failure. This has become known as feedback where something random leaves you achieving your goal nearly every time and taps into the desire to gamble. On the level of brain biochemistry, in the case of the pigeons they were getting a little dopamine hit every time they had a reward delivered of delicious pigeon pellets, but if there was a small element of uncertainty the dopamine hit was much bigger. The *Flappy Bird* game does just this and presents you with random challenges, as there is an unending stream of gaps to fly your bird through. Sometimes the gaps line up with each other and all you need do is fly in a straight line, an easy challenge, but then you get a gap at the bottom of the screen followed by one at the top. To succeed now you need to tap furiously to get the bird to fly from bottom to top and invariably you crash and have to start again. The game is then playing you, like Zeiler did the pigeons, and giving you a series of rewards followed by the occasional failure. Nguyen did not design *Flappy Bird* to be so addictive and was not thinking about how he could tap into these behavioural drivers. He just made a game that his experience as a game designer and as a human being told him would be a hit. By chance, he targeted enough of our addictive drivers to make his game a monstrous success.

Many of our interactions online provide us with the same goals, feedback and progress that *Flappy Bird* does. On top of that there are other things that have been identified as habit forming, in particular with respect to social media. When you add social interaction to the mix of habit-forming elements it seems to make everything more addictive. Take the innocuous little thumbs-up sign now found ubiquitously on social media platforms. Facebook launched the 'like button' in February 2009 and it has been a central part of the site ever since. It seems such a simple thing that Facebook innocuously describes as an 'easy way to let people know that you enjoy [a post] without leaving a comment'. But the like button is the crack cocaine of social media as, just like *Flappy Bird* did, it taps into several of the drivers of behavioural addiction, but unlike Nguyen's game is also synergized by social interaction. They didn't know it at the time, but they certainly do now. Mark Zuckerberg, the CEO and founder of Facebook, acknowledged that they needed to work out how to make the like button and other similar devices 'a force for good, not a force for bad'.

Back in South Korea, attempts are being made to work out how to tackle the rising tide of internet addiction. A study in 2017 looked at a small group of teenagers, half of whom were classified as being internet-addicted. They used a technique called magnetic resonance spectroscopy that allowed them to not only see inside the heads of the subjects but also to measure amounts of chemical transmitters. Professor Hyung Suk Seo at Korea University in Seoul, South Korea, reported that the internet-addicted kids had higher levels of a chemical called gamma aminobutyric acid, or GABA, and lower levels of glutamate in their brains. Both of these chemicals are what are known as neurotransmitters as they mediate passing signals between brain cells, although each of them is associated with opposite effects. The chemical GABA is associated with a reduced level of nerve transmission, while glutamate increases it. The net result of high GABA and low glutamate levels will be a slowing of brain signals which manifest as anxiety, depression and lethargy. We are not sure at this point if this is the cause or the effect of internet addiction. However, the good news is that when they took the internet addicts and put them through a course of cognitive behavioural therapy, adapted from one designed to treat gaming addicts, the raised GABA and depleted glutamate reset themselves to normal levels.

There are other treatments for internet addiction, most notably rehabilitation camps based on drug rehabilitation programmes. In South Korea, the camps take young internet addicts away from their homes, families and any internet connection. Then, alongside counselling, they try to show the sufferers how else they could fill their internet time, in some cases up to twenty hours a day. Crucial to these programmes is national recognition that there is a problem and the Korean government has put in place thousands of counsellors, hundred of programmes and even enacted the so-called 'Shutdown Law' that mandated that children under the age of sixteen should play no online games between midnight and 6 a.m. Asia very much leads the way in tackling internet addiction, not only in funding research and providing treatment but also in just recognizing that it exists.

We are now much more aware of how to turn an online activity into something capable of driving people towards behavioural addiction. Knowing these things has not stopped companies and individuals from creating addictive online content – far from it, in fact. 'Obsession engineers' actively seek to tap into these identified drivers to make their games and websites irresistible, hugely popular and thus very profitable. As a society, we need to ask how we cope with this new knowledge and how we exploit it and each other. Given that avoiding connection to the internet is becoming nearly impossible in the modern world, for people addicted to the internet, going cold-turkey and completely shunning the source of their addiction is not realistic. Better understanding of the biology of behavioural addiction will allow us to treat the symptoms, but many would argue that it is also time to bring the cause under control and that may mean curbing the freedoms of the internet.

THE QUIRKS OF THE HUMAN BEING

A sense of place

Try this with a mirror. Hold your arm out in front of you, point your index finger and then touch your nose. Presumably you bent your arm and your finger reached your nose without any issue. Now, close your eyes and repeat the action of touching your nose. Again, presumably you had no issue with this and once again touched your nose. Now, for the ultimate test, close your eyes and repeat the nose touching but this time stop your finger a couple of centimetres (an inch) away from the tip of your nose. How was that? When you open your eyes, your finger should be hovering just in front of the tip of your nose. Find somebody else to try this with and see how close they can get their finger to the tip of their nose without touching it, with their eyes open and with their eyes shut. My guess is that you and your companion nose-touching expert could accurately place a finger just in front of the tip of the nose with no problem, with or without your eyes open.

If you consider what you have just been doing, at first it seems quite extraordinary, as the usual assumption is that we perceive the world around us through our eyes and position our own body according to this information, taking further feedback from our eyes. While it is true that our ability to know where we are in three-dimensional space takes cues from what we see, there is a whole additional input to the system. You and I have an extraordinary ability to know at any instant where the various bits of our body are with respect to all the other bits. This sense is known as proprioception, and is the most important of the senses that fall outside of the usual five we were all taught in school. You can think of proprioception as an internal sense rather than an external one. The traditional five senses – sight, hearing, taste, smell and touch – are all external senses in that they all feed information about the environment outside of our bodies to our brain. Proprioception is internal and gives us a knowledge of the state of what's inside us.

Taken as a whole, proprioception gives us a complete picture of the relative position of all our limbs, the angle of our head and the twist or flex of our torso. The brain assembles this complex three-dimensional picture from a number of inputs. If we can see our limbs then that information is integrated and the fluid-filled system inside your ears also provides information about the orientation of your head relative to gravity and the surface of the earth. But none of this is strictly needed. If you ever find yourself in a zero-gravity situation in space, if you close your eyes you can still do the nose-touching proprioception test. Instead, we assemble the picture from an array of nerves each armed with special detector systems.

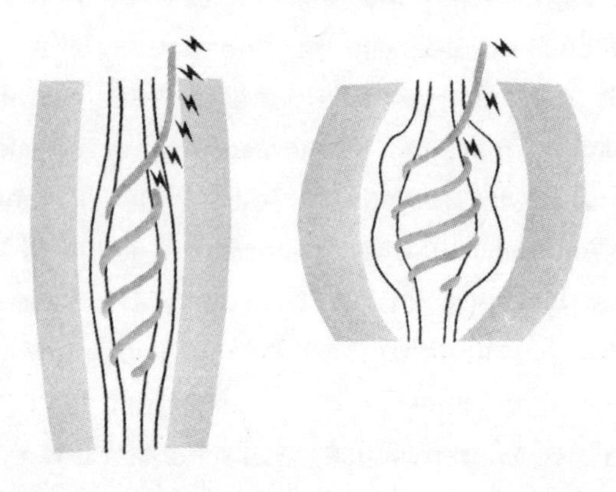

Muscle spindle fibre being stretched and squashed

The most important of these are the muscle spindle fibres. Inside each of your muscles are at least one and in most cases several bundles of fibres responsible for providing proprioception information. Muscles are made up of individual strands held together in a tough sheath that collectively can contract and apply force, usually to the bones to which they are connected (see page 124 for more on how muscles work). Tucked inside these sheathes of fibres are smaller muscle fibre bundles with their own protective wrapping. These smaller bundles are the spindle fibres. The ends of long nerves stretching from your spinal column wind their way into the muscle to wrap around the individual muscle fibres inside the spindle. When the muscle contracts or is stretched, the nerve endings are themselves stretched or squashed and they report this, feeding the brain with proprioception data on the length of the muscle but also the speed at which this is changing.

Other nerves provide information in a similar way from the specialized mechanoreceptors embedded in your joints and the Golgi tendon organs found, predictably, in the tendon tissue that connects muscles to bones. Taken together, these inputs give the brain sufficient understanding to form a picture of where all your limbs are, just from the angle of your joints, the length of your muscles and the strain on your tendons. The current theory is that, from the moment of your birth and possibly even before, your brain is building a picture of how these signals from your nerves correspond to the various bits of your body. This internal map of your body seems to exist, at least partially, from birth and is rapidly extended as the infant develops motor control over its limbs and movements. Proprioception in adults is something we mostly take for granted until we are unfortunate enough to lose some of this sense. There are pathological conditions that can lead to a loss of our sense of position, some genetic disorders or even an overdose of vitamin B6, for example. However, any sudden change in our body shape will put our internal map out of alignment with reality and most of us experience that at one point in our lives. The teenage growth spurt usually strikes at around fourteen years of age and averages a speed of about 10 cm (4 inches) increase in height a year for boys and slightly less for girls at 9 cm (3½ inches) a year. For many experiencing this rapid change in body proportions, it is accompanied by a period of awkward clumsiness caused by a mismatch between the teenager's body and their internal proprioception map of it. With time, the body relearns and adapts its internal map and hopefully the clumsiness passes.

There are a couple of instances in which a quirk of proprioception impinges on our lives. The classic medical knee reflex that I recall practising on my friends at school is an example of the proprioception system being fed incorrect data. The knee reflex test is used to check the part of your nervous system over which you have no control. It is traditionally administered by having a patient sit so their lower leg is free to swing and then gently tapping the skin, usually with a special rubber hammer, below the knee cap. You can just about do this on yourself if you cross one leg over the other and find a suitable device for the tapping, but it is much more convincing if you can find a helper to administer the gentle tap. What the tap is doing is stretching a tendon than runs from just below your knee cap, up and past your knee where it transmits that small stretch to the biceps femoris muscle that sits on top of your thigh. That small stretch of the muscle in turn elongates the spindle fibres inside the muscle. The result is that your body thinks that the lower leg and foot are moving backwards as that is what happens when the biceps femoris muscle is extended. Except that this is not what your body expects to be happening so it adapts and attempts to return the leg to the correct position by causing the muscle to contract, which in turn kicks the leg up. By sending a spurious signal to your proprioception system we can cause an aberrant movement.

The other place that proprioception has in popular culture is the classic test for drunkenness that used to be administered by police, before the advent of more reliable breathalyzer tests. The person suspected of being under the influence of alcohol was first asked to walk along a straight line on the ground and then to touch their nose. Both of these are tasks that rely on proprioception. Walking is, after all, a complex activity of foot and leg coordination that we rarely observe ourselves doing, not least because it is happening partially behind and beneath our torso, where we can't even see. Touching your nose, which we tried at the start of the chapter, depends entirely on proprioception as well. Since alcohol has a debilitating effect on our sense of proprioception, the simple nose-touch and line-walking tests are an effective, if crude, way to measure inebriation.

Possibly one of the most remarkable things about proprioception is that it can be trained to include things that are not part of your body. This is an extension of the idea of muscle memory, that by practising a physical action over, and over, and over again you can make that action essentially instinctive and thus free up your brain's processing power for other activities. The playing of a musical instrument is a classic example of this, as a proficient musician does not think about where their fingers need to be individually but has developed a muscle memory that helps them concentrate on the music they are playing, rather than the physical act of pressing keys or plucking strings. This clearly is a process that goes hand-in-hand with proprioception, but in some cases has extended it. A proficient violin player, for example, will know exactly where the bow of their violin is on the instrument without needing to look at it, and they can keep their eyes on the musical score while playing. This is known by the scientific community as extended physiological proprioception and helps explain not only phantom limb syndrome, but also some artistic activities.

The broken escalator phenomenon is an example of proprioception going wrong. When you step onto a stationary escalator, your experience tells you to expect a sudden change in momentum and speed. To compensate, your brain anticipates but fails to receive the appropriate proprioceptive response from your body. The result is a peculiar feeling of imbalance or, in some cases, nausea.

Take, for example, blind contour drawing, a technique where the student is told to draw a picture of what they see without looking at the paper. The idea is that once you have placed your pencil on the drawing paper you do not take it off. You then focus on the object to be drawn and imagine the pencil as an extension of your own hand and then mirror the movement of your eye as it slowly traces the contour of the subject. At least that is the theory. In practice your first attempt will be rubbish, as will your second and third and fourth, but eventually you will begin to get the hang of it. Your proprioception map of yourself will be expanded and will now include a pencil and you will teach yourself to mirror the movement of your eyes' focus onto your pencil tip.

While I would love to be able to blind contour draw and create wonderful pictures while never looking at the paper I am working on, I suspect that if I want a demonstration of the power of proprioception the 'not quite touching your nose with your eyes closed' test is probably a lot easier to master and perform.

Bacteria make you human

Over the last ten years or so, a new idea has gradually been taking hold and turning the tables on much of our understanding of human biology. Before this revelation it was believed that your body was made up of a number of organs big and large that interacted together to create a human being. External influences could certainly impact the workings of the body, often in a negative way, but when it came to the hormones that influence your mental well-being, your sporting ability or your appetite, the control was all inside you and linked to your own organs. What has become apparent is that there is another source of control that is not strictly human – the bacteria living in you.

The human body is made up of some 37 trillion cells, which is an unfeasibly large number (see page 132 for more on big numbers). For a long time, it has been known that there are at least the same number of bacteria living in you. Most of these bacteria can be found in your intestines, but they are also all over your skin. The figure you see often quoted of 100 trillion bacteria inside a person has recently been downgraded somewhat to a mere 40 trillion bacteria, a number about 10 per cent greater than our body cell count. So, technically, if you consider the number of cells, you and I are more bacteria than we are *Homo sapiens*. The weight of these bacteria you are carting around with you all the time amounts to about two kilograms, or roughly 2 per cent of your body weight. If you consider these bacteria as a working organ within your body then that makes them the heaviest of human organs. However, why would you consider them an organ if they are just hitching a ride in your guts, feasting on your food and doing nothing? That is what science believed twenty or thirty years ago. Then work was done that began to show a different story. It has become apparent that the bacteria in you, now called your gut microbiota, have a vital role in keeping us healthy and

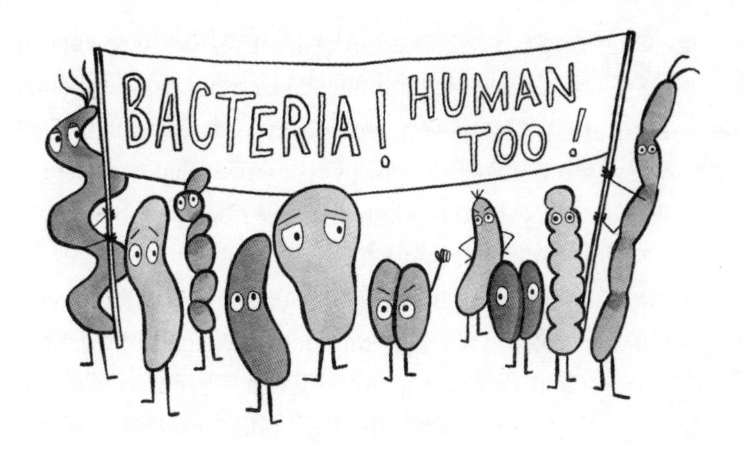

modifying our moods, primarily by the release of hormones.

The work originally started by looking at how mice with no bacteria behaved when compared to those that had a normal bacterial load. Creating a sterile mouse, by which I mean free from bacteria rather than infertile, is no mean feat. It is not sufficient to take newly born mice and transfer them to a super-clean environment, as any newborn baby creature will pick up bacteria as part of the process of being born. Creating bacteria-free mice is a laborious task involving antibiotics and repeated generations to slowly but surely eliminate all bacteria from mice that must then be raised in a sterile and bacteria-free environment. The mice that result from this process are not generally very healthy. They have eating problems, perhaps unsurprisingly, but they are also not sociable and when this was investigated it became clear that their brains had not developed properly. The missing bacteria were influencing the mice on a fundamental, developmental level.

Since then, numerous similar studies have shown a variety of effects on mice. Taking this work from the carefully controlled laboratory experiments, with their sterile animals, and into humans is much trickier. We cannot create sterile humans the way they did with mice so we are forced to do what most medical studies do: use double-blind controlled studies. In order to generate proof that a treatment is having an effect, the accepted procedure is to gather a large group of volunteers, selecting people based on the profile you want to mimic in the actual population of interest, and then to treat some with the treatment or test, while the others are given a placebo or dummy treatment (see page 162 for more on the placebo effect). The double-blind part of the name is because the trial needs to be arranged so that neither the patients nor the researchers know who is being given the real treatment and who gets the placebo. That way, unconscious bias on the part of the researchers cannot influence their analysis and the patients are not biased in their reporting of what happens. You then need a big enough sample to be statistically valid. It is no use doing the trial on a handful of people as the result you get could just be a coincidence. Depending on how small or nebulous the effect is, you may need hundreds or thousands of test subjects. So, while double-blind trials are the gold standard to test a hypothesis about human biology, they are really hard work and very expensive. Which is why so many small-scale studies take place.

Such as, for example, the study carried out by Professors Ted Dinan and John Cryan at University College Cork in the Republic of Ireland. This professorial pair had previously put sterile mice under stress and analyzed how they coped with and without healthy gut flora. When they transitioned to human subjects, though, they needed to control some of the variables in the experiment and decided to use just a single gut bacterium called *Lactobacillus rhamnosus*, which had helped the stressed mice relax. Initial results showed no effect in humans. Just going to show that not all tests transfer between species, they then tried a different bacterium called *Bifidobacterium longum 1714* and this seemed to be doing something. They only had twenty-two people in their trial, but when half were given a daily pill containing the bacterium, they showed reduced levels of stress hormones and reported less anxiety compared to the other half who had the placebo. It is a tiny sample, but enough for the bacteria to be marketed as a probiotic dietary supplement called Bimuno, although there are now myriad brands. In an attempt to get to the bottom of the validity of this treatment, the BBC commissioned and televized a further trial that looked at how or if Bimuno helped with insomnia. The trial was funded by the manufacturer of the product and they were delighted with the positive results that led to a huge spike in sales. Ironically, the trial done for the television was even smaller, with only one participant. A bigger trial is clearly needed, but the manufacturer already has a market for the product and thus has no need or motivation for funding a larger study.

Then there is the story of Lauren Peterson working in Connecticut in the USA, who apparently changed her sporting ability through a faecal transplant. One of the consequences of all this work on the bacteria that call us home is the realization that if you can change those bacteria you can have an effect; for example, in curing insomnia. An extreme version of this, then, is to not only take a pill containing bacteria but to insert a new culture of bacteria into your intestines. There is no delicate way to put this and a poop transplant can be seen as at best a suppository of donor poop or as a reverse enema. Peterson was working as a PhD student on gut microbiology and knew that she had a particularly deficient set of bacteria in her intestines. As a child she had contracted Lyme disease, an unpleasant bacterial infection carried by ticks, common in New England in the USA. The disease causes debilitative bouts of joint pain and tiredness that can keep recurring for many years after the initial symptoms pass. As a result of spending much of her childhood being treated for the disease with antibiotics, Peterson had a severely depleted microbiota and she knew this because she had submitted samples to the American Gut Project on which she had come to work. As a side project, she was working on another area that she was passionate about: cycling. She collected stool samples from cyclists of all types. She had samples from casual cyclists, people who trained a bit for amateur races, like herself, and then thirty-five samples from elite cyclists. She looked to see if there was any commonality between their gut bacteria. Two things stood out. First was a genus of bacteria called *Prevotella*. The more you cycled competitively, the more chance that you had *Prevotella* in your poop. Only about 10 per cent of the casual cyclists and half the amateurs carried this bacterium, but all the elite cyclists had it. The second bacterium that she singled out wasn't even a bacterium, but a member of the peculiar kingdom of Archaea that look a lot like bacteria yet have a much more ancient origin. Members of this kingdom of life tend to be found only in the most extreme environments; in sulphurous pools of boiling water, or in the deep oceans. In this case, the Archaea species *Methanobrevibacter smithii* has found a niche where it can survive on the carbon dioxide and hydrogen produced as a waste product by other regular bacteria in the gut. Most of the elite cyclists had this peculiar organism growing in their guts. Thc working hypothesis suggested by Peterson is that since this organism digests the waste products created by other bacteria, it prevents the build of this waste which is toxic to the bacteria that produce it. Consequently, the bacteria can go about their business helping digest your food for longer and you extract more energy from the food you eat. It's possible that *Methanobrevibacter smithii* makes digestion more efficient. Given that it appeared to Peterson that, firstly, the elite cyclists had a particular array of gut bacteria, that secondly she did not have this bacterial array, and thirdly she was keen on becoming a better cyclist, she gave herself a self-administered faecal transplant from a donor elite cyclist. She claims the result was remarkable. She went from training a few days a week to training every day, with an energy she had never before experienced. Concrete proof of her cycling success was forthcoming as she began competing and even winning in professional endurance races. Can we put all of her success down to her poop transplant from an elite cyclist? Sadly no, without a control and double-blind trial it is just an anecdote. Other factors in Peterson's life will have contributed to her success. I note that the life-changing transplant happened at the end of her PhD, a period that anyone who has been through it can attest to its ability to change your life. Her work has of course opened up a new area of research, but also the worrying possibility of poop doping becoming an issue in the future for sport.

Bacteria are not just found in our guts, they are all over us. The Belly Button Biodiversity project, run from North Carolina State University in the USA, has found over 2,000 different varieties of bacteria, with an average of sixty-seven types in each belly button.

It is clear now that the bacteria that live on and inside us are more than just hitching a ride. While not all bacteria are good guys, even with pathogenic bacteria our own microbiota can apparently help out. The presence of *Bacteroides fragilis* in your gut can, so long as it stays there, help your immune system combat other infectious bacteria. Even the bacteria on your skin have a role, with evidence pointing to them helping fight infections of parasites. So, do we risk damaging this only recently discovered ecosystem of helpful bacteria if we are too cavalier with anti-bacterial agents and antibiotics? Should we be exposing ourselves and our children to more bacterial diversity? There is a growing body of evidence that the obesity epidemics in Western cultures such as the USA may have, at least partially, a link to the bacteria in the gut. How, then, can you establish a healthy gut microbiota and maintain it? The secret turns out to be no secret at all and a healthy balance of foods, with a preponderance of vegetables and fruits giving your gut bacteria the best chance to develop a rich and healthy diversity. However, there is only so much that can be done, as there appears to be a genetic dimension to which bacteria thrive inside you and which do not. One thing is for certain, though: the use of antibiotics makes a mess of the bacterial community inside you. While it may help clear that infection, we now have yet another reason, on top of the rise of resistant bacteria, to give pause to where and when we use these drugs. There is always the possibility of resetting your gut microbiota with a poop transplant. Since the very best transplant would always be from a source that you know is guaranteed to be compatible, and that means yourself, some researchers have gone to the extent of keeping in their freezer a stool sample from all their family members, just in case one of them needs a faecal transplant and a top up of the good bacteria.

Dementia and your teeth

In 1906 the German psychiatrist Alois Alzheimer published a report on 'a peculiar disease of the cerebral cortex' that he had identified in a single patient. The disease was a form of dementia and over the next five years another eleven cases were described in the literature. Initially the classification of the disease described by Alzheimer was restricted to patients under the age of sixty-five, but in 1977 the consensus among psychiatrists changed and it was realized that the disease was much more prevalent in older people. These days, Alzheimer's disease accounts for about 70 per cent of all dementia cases and in 2015 was listed as the cause of death for forty-seven million people, the fifth biggest cause of death globally (see page 55 for more on what causes death). This death toll is estimated to double every twenty years, mostly due to the ever-ageing population of the world. While it is a relatively recently discovered disease, we have been aware of it now for over 100 years. So, it may come as a surprise that the cause of Alzheimer's disease is still unknown.

What was known from early in its history was what could be seen in the brains of patients who had died suffering from the disease. The tissue of their brains was shrunken and filled with peculiar structures known as plaques and tangles. The plaques formed outside the cells of the brain, the neurones, while the tangles were found inside the neurones. Down a microscope, the effect of the disease was to give tissue samples a blotchy appearance from the plaques and the cells to become lumpy and wrinkled from the tangles. A major breakthrough came in 1984, when a group at the University of California isolated and identified the substance from which the plaques were made. The culprit was a protein given the name beta-amyloid and within a few years of its discovery a hypothesis had been put forward that pointed to this protein as the causative agent of the disease, with the assumption that as the plaques begin to clog up the brain the characteristic symptoms of Alzheimer's disease appear.

Armed with a hypothesis and a pressing need for some sort of medication to help the sufferers of this widespread and common disease, researchers began looking for ways to treat it. Governments and drug companies invested huge quantities of money in the search for a treatment. So far, the search has been very disappointing. If you follow the scientific press, you will have seen regular announcements of breakthroughs that herald a way forward, only for those breakthroughs never to be heard of again. If you dig into the literature to follow up on these discoveries, invariably what started as a promising experiment in small-scale trials turns out to have failed at the next hurdle. The best treatments we have come up with so far have had only very marginal benefits to sufferers. Between 1998 and 2014, some 124 different drugs were trialled to counter Alzheimer's disease. None of these have passed all the tests and become commercially available.

When anyone finds themselves banging their head on the walls of an intractable problem they will eventually stop banging, pause and consider if there may be another way.

In the case of Alzheimer's disease, people began to question the initial assumptions. Was the amyloid hypothesis wrong? All people suffering from Alzheimer's disease have amyloid plaques but that does not mean the plaques are causing the disease. The plaques could be a symptom of the disease instead. Which is an especially convincing idea when you consider the so-called super-agers. A team of scientists based in Chicago in the USA were studying a group of people over ninety who did not have Alzheimer's disease. In fact, they displayed almost the opposite symptoms and on tests had memory and understanding scores equivalent to people in their fifties. There was no obvious cause for the super powers they displayed, with no lifestyle clues as to how they retained their mental agility. Investigating further with brain scans, it became clear that they did not display the normal shrinkage of the brain that somebody in their nineties usually does. In an attempt to understand a bit more about what made the super-agers so super, they looked at brain samples donated to science after death and found something completely unexpected. The brains of super-agers contained the telltale plaques and tangles associated with Alzheimer's disease. In fact, the oldest of the super-agers displayed the full range of pathological signs that would lead a clinician studying such a sample to diagnose Alzheimer's disease. And yet these patients showed no outward symptoms of the disease. We still don't know why super-agers retain their mental agility so late in life, but it does give credence to the idea that the amyloid hypothesis may be barking up the wrong tree.

To identify the cause of a disease there is an established list to check, known as Koch's postulates, after the eponymous, nineteenth-century German microbiologist. First you must find the disease-causing organism in all people suffering from the disease. You must then identify and isolate this organism from somebody with the disease. Once isolated, the disease-causing organism is then reintroduced into a test subject, who should then display the symptoms of the disease. Finally, you go through the second step again and isolate the disease organism from your now sick test subject and prove that it is the same as the one originally found. At this point, according to scientific dogma, you are ready to claim you have identified what causes the disease, except of course that life, and particularly biology, is never that simple.

In 2011, biologists studying the epidemiology of Alzheimer's disease noted that people with fewer teeth were slightly more likely to get the disease. Investigating further, they discovered that the link seemed to be specifically between Alzheimer's disease and gum disease, which is caused by a bacterium living in your mouth called *Porphyromonas gingivalis*. This bacterium, if left unchecked, grows down between the surface of the tooth and the gum, eventually causing bone loss and the tooth to fall out. It's a common complaint, with about 70 per cent of those over sixty-five having the condition in some form or another, usually in a minor way. Treatment for gum disease is relatively straightforward and exactly what you expect for something within the domain of the dentist: brush your teeth carefully, use dental floss or dental sticks, don't have sugary drinks between meals and maintain a good oral hygiene routine. Which is, of course, easier said than done.

Evidence has begun to mount up, though, that the link between Alzheimer's disease and gum disease is not some odd chance correlation and there may indeed be a causal link. Mice are the subject of choice for Alzheimer's disease research, as a special genetically modified breed has been created that suffers from something very similar to the human version of the disease. So far, the gum disease bacteria has been found in the same location as plaques in the brains of mice suffering from mousey Alzheimer's disease. Mice specifically given gum disease end up with Alzheimer-like conditions and develop amyloid plaques. The science is trickier with human subjects. The first step taken was to find a good way to identify when a tissue sample taken from a brain contained the gum disease bacterium. Fortunately, *Porphyromonas gingivalis* produces some unique, if unpleasant, protein enzymes that it secretes into the surrounding tissue. These gingipains help the bacteria invade the gum tissue by breaking down other proteins on the surface of the healthy gum cells. When scientists from Lancashire in the UK looked for the gingipains in brain tissue, they were found in the same places as the amyloid plaques. Experiments have also shown that if you introduce the gum disease bacterium into a mouse's brain you get amyloid plaques forming within a day. If we jump back to Koch's postulates, it seems we are most of the way to proving the disease-causing agent is the gum disease bacterium.

An organism, *Porphyromonas gingivalis*, has been found in brains from people who suffered from Alzheimer's disease. It has been identified and reintroduced to an unaffected brain, caused the same symptoms and then been identified at the site of the symptoms. That is all four of Koch's postulates satisfied, admittedly spread across several different research groups and using mouse and human studies. But this is not a simple case and many researchers are less convinced.

The most common bacteria in your mouth is not *Porphyromonas gingivalis* but a different species, called *Streptococcus mutans*, which causes tooth decay. Its genes allow it to stick efficiently to the surface of teeth, forming a biofilm, and then digest sugars to create lactic acid. The acid demineralizes tooth enamel, leaving tiny pores that harbour more bacteria until a cavity appears.

Firstly, we don't know how the gum disease bacteria could possibly get from the mouth to the brain. You can concoct a scenario in which the bacteria enter your blood stream when you vigorously brush your teeth. Once in your blood, the bacteria would then need to pass through the blood-brain barrier, a specialized filtration system that stops undesirable matter such as bacteria leaving your blood and getting into your brain. We know that *Porphyromonas gingivalis* can invade white blood cells, so it is possible that it does this to sneak its way through the blood-brain barrier. On top of that, there is a significant genetic factor in susceptibility to Alzheimer's disease and we don't know how that plays into this bacterial hypothesis. The start of the story may be that the most important gene that associates with Alzheimer's disease is one that makes the ApoE protein, and the gingipains are very good at attacking this protein. Finally, there is the small issue of how the bacterium causes the disease once it gets into the brain. It is one thing to work out how they get there but if there is no obvious mechanism for the disease then despite Koch's postulates being fulfilled we are still none the wiser. At the moment the likely candidate for this last piece of puzzle is that the gum disease bacterium sets off a defence mechanism in the brain that smothers the bacteria with amyloid plaques. Then, if your genetics make you susceptible, the defence mechanism gets a bit carried away and ends up killing brain cells, as well as bacterial ones.

So, does that mean we will have a cure for Alzheimer's disease any time soon? I am guardedly hopeful. The amyloid hypothesis has for decades now not yielded any great breakthroughs and this new approach may bring relief for sufferers and possibly one day even a cure. Researchers in Australia are already working on a vaccine for *Porphyromonas gingivalis* and international drug companies are starting to turn their eyes to possible ways to target the bacterium. If it does turn out that gum disease bacteria are at the root of Alzheimer's disease, a side effect of any treatment may also be an improvement of dental hygiene.

Exercise is a pain

I am a fair-weather runner. Each spring since I took up running about six years ago, I dust off my running shoes, quite literally, and get myself back out doing some exercise. I like to go for runs outside, usually along my local river in the nature reserve, but if I'm away for work I can often find myself a park to enjoy. Now, don't get me wrong, I am no athlete. I plod along for half an hour or so and try to get close to the five-kilometre mark – I prefer to run kilometres, as that sounds much better than just three-and-a-bit miles. When winter sets in I attempt to keep going, donning extra layers and gloves, but my heart is just not in it. One year I even signed up for a gym and tried to continue through the dark and dank months, but pounding a treadmill, going nowhere and only seeing a white painted wall was not my idea of fun. So, the running comes to a stop for three months. When spring once more arrives and I head out for the first run of the year, I always end up with a case of delayed onset muscle soreness or, as I discovered recently, the effect can be conveniently conveyed for social media purposes as #MajorDOMS.

The symptoms of delayed onset muscle soreness, and I am going to start abbreviating it to DOMS, are simple and happen with almost all exercise. Immediately after your exercise you may feel a bit sore or tired, but this is not DOMS. As the name suggests, the soreness from DOMS is delayed and usually hits me the next day or occasionally two days after. The affected muscles are painful when extended and even tender when touched. It lasts for about a day, or maybe a bit more if the exercise was particularly extreme (#MajorDOMS). What is peculiar about DOMS is that the second run I go for in the year I may get a little bit of soreness, but by the third it is gone. This year, as I write, it is the start of spring and the running has just begun again. I am four runs in. I was sore two days after the first run and was hobbling about for a day, following the second run I noticed it a bit but run three and four yielded no DOMS. How then do we account for the peculiar nature of this pain that appears two days after the exercise but only the first time we do that exercise?

The story of how muscles work and how we came to that understanding is a classic of cell biology that is rarely told. Which is a shame, as one of the central characters was a brilliant woman called Jean Hanson. Just after the war at the start of the 1950s, Hanson, a biophysicist who had just finished her PhD, took a year-long sabbatical at the Massachusetts Institute of Technology in the USA. There she met another postdoctoral researcher called Hugh Huxley and together they constructed a theory of how muscles worked. By this time it had been established that muscle consisted primarily of two proteins: one called myosin and the second called actin. We knew that myosin broke down the energy-storage chemical used within cells called adenosine triphosphate or ATP, and that peculiar things happened to mixtures of the purified proteins. If you put pure myosin and pure actin in water together they create a gloopy gel, until you add ATP when it thins, turning back into a liquid, for a while at least. Combined with careful microscopical work that showed muscle to be made up of alternating light and dark bands, Hanson and Huxley put forward the sliding filament theory of how muscles work. They could see that the actin was forming long filaments that were aligned parallel to each other. Imagine them as the fingers on your left hand. The myosin also seemed to organize itself into filaments, once again aligned in parallel arrays but lying interspersed with the actin protein filaments. In this case, imagine the myosin filaments are the fingers of your right hand, and now interlace your fingers so that just the first joints of each finger overlap. When ATP, the universal cellular energy source, was added to this complex arrangement of actin and myosin, one set of filaments would slide over the other. In my finger analogy you now slide your fingers together and, this is the important bit, your hands get closer together. Hanson and Huxley proposed that this was what was happening with the actin and myosin. Each muscle they could see in the microscope was made up of millions of these little arrays of parallel filaments, which when flooded with an energy source could slide together, each shortening by a tiny amount. Overall, the effect is that the muscle contracts significantly along its length.

We now know much more about how the whole process works together to give rise to muscle contraction. The actin protein was found to be what is known as a globular protein. On its own, a single protein is just a roughly spherical blob, but put a whole load near each other and they start to self-organize into long, winding chains. It turns out that these actin chains, or filaments, are the scaffold structure within muscles and all cells. Myosin, on the other hand, is a much more interesting protein molecule. It has two parts to its structure, a long wiggly tail and a bulbous head at one end of the tail. Just like the actin, myosin will self-organize into filaments by wrapping its tails together, resulting in a filament with the bulbous heads poking out along its length. Within muscle tissue, the actin is organized into tiny units of parallel fibres all connected together at one end along what are called Z-disks. You then have two sets of these arrays facing each other with double-ended myosin filaments joining the two. When ATP comes along, the heads of the myosin molecules grab hold of the actin scaffold bars, flex down towards the midpoint and ratchet the myosin filaments towards the Z-disks. Since this is happening at both ends of the myosin filament, the Z-disks move closer and the muscle shortens. It is a beautiful piece of molecular scale engineering, all done with proteins. It is also the reason behind delayed onset muscle soreness.

Every time you use a muscle, countless millions of myosin molecules are ratcheting back and forth and frantically crawling along actin filaments. If you keep working any one muscle too much, a few things happen. Firstly, inside the muscle fibres you start to use up the molecular energy store ATP at such a rapid rate that waste products from its creation begin to build up faster than our muscle cells can get rid of them. The waste product is lactic acid and it is this that makes your muscles start to ache and eventually give the burning sensation when you go to extremes, but it is quickly cleared away. Resting the muscle for just ten to thirty minutes will completely remove any excess lactic acid and you are ready to go again with the pain gone. What you will not have noticed, though, is if you have caused enough damage to give rise to the delayed onset variety of muscle pain.

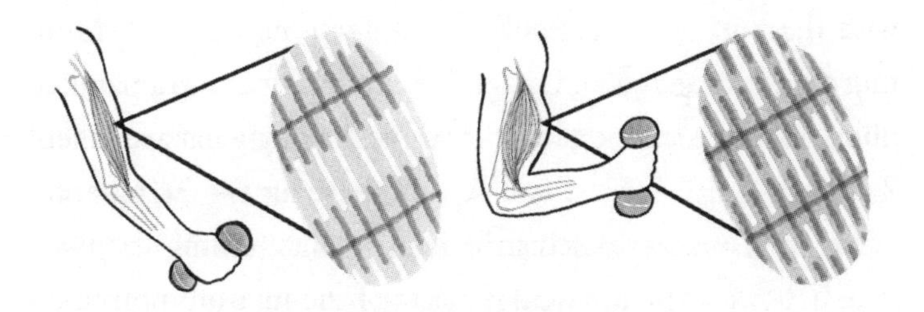

Actin and myosin filaments working together to contract a muscle

Delayed onset muscle soreness is associated with a particular type of exercise known to sport scientists as eccentric, as opposed to concentric, exercise. It is never helpful to the layperson when science chooses to use words with perfectly respectable ordinary meanings to convey something completely unrelated and very specific. To explain eccentric and concentric exercises it is easiest to give an example. Imagine an exercise that involves repeatedly lifting an object held in your hand upwards and downwards, closer to and then away from your chest. This is the classic dumb-bell curl that works on your biceps muscle, but it need not be a dumb-bell in your hand; it could just as easily be a book, a baby or a pint of beer, although probably not all at the same time. As you lift the object, your biceps muscle at the front of your upper arm shortens or contracts and the exercise is said to be concentric. When you then lower your arm you still need to use your biceps to control the descent of the object, extending the muscle, and this is when you are doing eccentric exercise. You only get delayed onset pain from eccentric exercises, when you are controlling the elongation or extension of a muscle.

What happens when you perform this kind of exercise for the first time, or after a long break, is still being discussed and researched. However, the best candidate for the explanation currently is that by putting force on your muscle as you allow it to extend you cause damage to the little subunits of myosin and actin filaments. Recall that the actin is joined together and each unit is glued to the neighbouring one by the Z-disc. Eccentric exercise seems to put this structure under a lot of force which can cause it to start to split apart, giving rise to micro-tears in the muscle. You can't see anything with the naked eye, but on a microscopic level the muscle is literally tearing itself apart. Once micro-tears have formed and you stop the exercise you feel fine with no pain. But your muscles are full of damaged tissue and that needs to be fixed. Very quickly your immune system kicks into action as the damage will have released a whole slew of proteins into the fluid between cells and consequently your blood. White blood cells floating about in the blood pick up on these leaked proteins that aren't supposed to be there and signal for an immune response by releasing chemicals such as histamine and serotonin. These chemicals make all the local blood vessels expand and increase the blood flow to the area. Fluid begins to flow into the damaged muscle fibres, making them swell slightly. Following close on the heels of this fluid are white blood cells called neutrophils that wiggle their way into the muscle tissue and begin hoovering up the damaged proteins. The whole process is what a medic would call inflammation and is a normal response to any damage in your body, no matter the cause.

As can be seen, delayed onset muscle soreness is a complicated multi-step process, so it is no surprise that inflammation takes a little while to get going, which is what puts the delay into its name. Twenty-four hours after you exercised and caused lots of micro-tears to open up in the muscle, inflammation sets in and that is when the pain starts. Discomfort caused by DOMS is not usually noticed when you are resting the muscle, only when you try to use it. As soon as this happens, you try to contract the muscle but the inflammation adds an extra squeeze and that in turn activates pressure-sensitive receptors in the muscle and you feel pain. After another day or two the damage to the muscle is repaired, the inflammation recedes and any pain does, too. The great thing about this repair process is that it conveys upon you the repeated-bout effect and immunity from pain the next time you go for a run or take exercise.

A slightly different version of the protein myosin is responsible for ferrying cargos around inside your cells. In order to move a load, it is hitched to a myosin molecule and this is attached to part of the actin filament network within the cell. It's as if each cell has a rail network and the myosin acts as the locomotive.

The repeated-bout effect is what stopped me from suffering delayed onset muscle soreness after my third and fourth and hopefully all the subsequent runs this year. Again, the exact mechanism is still being investigated but it looks like the repair process instigated after you inflict micro-tears on your muscle not only repairs but enhances your muscle. The number of myosin and actin subunits within the muscle increase during the repair process. Your muscle is now ever so slightly longer and able to perform the harmful eccentric exercise without any damage occurring. This effect, caused by one episode of delayed onset muscle soreness, can last for up to six months. Although in my experience it is less than that, only a month or so, which explains why each spring I have to suffer anew. My first run of the year inevitably ends with pain two days later but the inflammation and healing that takes place then sets me up for the rest of the year. Even if I miss out on a few of my regular runs, and I will admit to the occasional four-week gap, the next time I run I don't suffer any muscle soreness after the exercise. My muscles will have been adapted by the inflammation and repair process to lengthen and I can run the same distance with no ill effects. Better yet, as long as I maintain the intensity of my exercise I could in principle run for longer and do more exercise. While this is a theoretical possibility, for me it remains an unlikely one.

As long as you maintain an occasional session of exercise, the repeated-bout effect will stay with you. If, as I do, you allow yourself to lapse for a significant chunk of time, a process as yet unidentified will trim away all the extra myosin and actin subunits and you are back where you started. What, then, can understanding this science do to help us avoid the pain of delayed onset muscle soreness? Clearly the most obvious way is to hang on to the repeated-bout protection by carrying on with the exercise. You can reduce the frequency and you can reduce the intensity, but you just need to keep doing it. If, like me, you have tried that and it didn't work and there is going to be a long period of inaction, then rely on the quirk of biology that means you can get the same repeated-bout effect from just a short burst of exercise.

If I was to take my own advice, my first few runs of the year would all be really short ones and I would slowly build up to my full distance. Unfortunately, I am also far too competitive with myself and determined to prove that I can still do my usual run. Consequently, psychology gets in the way of sound sports science and two days after that first run each year I'm hobbling, stricken with delayed onset muscle soreness.

Big numbers are a bother

If I ask you to picture in your mind five people gathered together into a tiny crowd, you won't have any problem with the challenge (see page 178 for more on crowds). Twenty-five people and it gets a little bit trickier. What about 100 people or 500? Can you visualize this number of people standing in a field? Chances are, you are starting to get only a vague idea of what this might look like. If we further increase the numbers it all starts to fall apart. What does a thousand people look like, or a million or a billion? Admittedly you may also be starting to wonder about the capacity of this field, but the point stands. On the other hand, you would easily be able to say that a million people was a lot more than a thousand. Why is it that big numbers are so hard to grasp and yet we can still manipulate them?

The latest research seems to indicate that our brains are wired to deal with big numbers differently than small ones. It's not just people that have this ability, though. It has been shown in other primates and also small fish. The guppy is an easy test subject for an animal behaviourist to work with: it reproduces quickly, is simple to keep and very small. Guppies are also naturally gregarious creatures and will always try to gather together with the biggest group in their vicinity. Newborn guppies innately have this tendency and that gave Italian scientist Christian Agrillo, working at Padua University, a way to test numeracy in fish. He set up a tank for the guppies made of three compartments. At opposite sides of a main tank were two small spaces that could be stocked with a varied number of adult fish, creating a little shoal or crowd on each side. In the centre of the main tank between these two crowd tanks Agrillo would place a single young fish that had been reared with just one other fish for companionship. When the two crowd tanks were stocked with small numbers of fish up to a maximum of five, the young fish in the main tank would unerringly move towards the side that had the most fish and the bigger shoal. It could differentiate between a difference of just a single fish in the two crowd tanks. Conversely, if you increase the number of fish in each crowd to numbers up to twenty-four, the juvenile guppy was not so good at counting. Even with numbers just a bit bigger the guppy was less accurate. When given a choice between a shoal of six and a shoal of eight, the guppy was much worse at picking the bigger crowd, but still could pick the difference if the proportion was big enough. Slightly spookily, this work can be exactly mirrored with human test subjects, although rather than making a person choose between crowds they had them look briefly at sequential images of dotty patterns on a computer screen and then pick which image had the most dots. This work has been taken to mean that we have two number-recognition systems in our brains. The first is good for smaller numbers on a scale we are familiar with on a day-to-day basis: the number of digits you have on one hand, the number of glasses to set on a table or the number of pockets to check for your keys. This system gives us precise information about the number and allows us to process specific mathematical similarities or differences. We can instantly see that four fingers are not the same as five and which number is the greater. Then there is the system for grasping bigger numbers that for guppies kicks in above five. This gives the viewer of the objects a more intuitive, but less precise or mathematical understanding of the number, which is why it is only reliable for big differences and gets close variations wrong.

Now, you may be thinking that in the case of human test subjects they are just counting the number of dots and getting a precise answer. Except that in multiple studies the image has been flashed so quickly you don't have time for that sort of mental processing. Which certainly mirrors my own experience of how my brain is aware of how many objects I see. With small numbers I can look and just know that there were five biscuits on the plate before my children came in the room and now there are four. I don't count the objects but still know the number when it is a small number.

The numbers billion and trillion have completely different mathematical meanings depending on where you are. In the USA, the UK, Russia, North Africa and Australia, a billion is a thousand million and a trillion is a thousand billion. However, in Continental Europe, southern Africa and South America – but not Brazil – a billion is a million million and a trillion is a million billion.

These two number-recognition systems would also appear to be located in different parts of the brain. Experiments done with monkeys have demonstrated that when they are shown images of between one and five objects there is activity in a part of the brain called the intraparietal sulcus, a groove running along the bit of the brain just behind the very top of your head. In this area, some groups of nerve cells seem to be switched on in a linear pattern where the more objects shown to the monkey the more cells are activated. Other regions within this same area were linked to specific numbers. So, a patch of cells was activated when three objects were shown and a different patch with four objects. The intraparietal sulcus also looks like it is the area associated with mathematical ability. For those people who are diagnosed as having dyscalculia, any sort of mathematics is a huge uphill challenge no matter how hard they try. Within the brains of these people there is a reduction of brain tissue in the intraparietal sulcus. What makes this part of the brain crucial for our appreciation of big numbers is that they don't seem to cause any particular parts to become active. The intraparietal sulcus is only turned on by small numbers, the bigger ones leave it cold. How we process these bigger numbers is not known and possibly requires much more complex estimations that are not hard-wired in the same way as for small numbers.

The consequence of this dual system has some significant ramifications. While we can appreciate and immediately grasp the smaller numbers that are, after all, what we evolved seeing around us, when it comes to bigger numbers we have no reference points and it starts to change how we make decisions. Loran Nordgren of Northwestern University, Illinois, in the USA, set up a study in which he asked volunteers to decide on the punishment that should be meted out to perpetrators of hypothetical crimes. In each experiment the volunteers were split into two groups and given one of two situations to judge. The first experiment asked the volunteers to decide the length of jail term, from one to ten years, for a fraudster who they were told had either defrauded three or thirty victims. It turned out that the more victims the criminal had, the lower the sentence the volunteers handed out. Nordgren tried a range of different scenarios and it was always the same. When there were very few victims, the sentence was higher. The only way to equalize the sentences handed out was to show the volunteers a picture of one of the large number of victims and give the person a name, thus creating a number of people that the volunteers could empathize with, that is one person, and the sentencing ceased to be about a big number.

In the twenty-first century, the world we live in is surrounded by gargantuan numbers. We are at the beginning of the era of so-called big data, where only computers really have any chance of getting to grips with numbers that easily range into the billons and trillions. These could be interactions on social media, viral advertising, your genetic sequencing data or information about the weather from thousands of weather stations, which all produce huge numbers that are beyond our primate brains to really comprehend intuitively. Given the importance of these data on our democratic processes, we can only hope that those with the power to influence take care to come up with strategies to make sense of really big numbers.

Hairy beasts

Compared to a gorilla you may think of yourself as not very hairy – we are, after all, known as the naked apes for our distinct lack of the stuff. However, if you compare the number of hair follicles on an ape and a human we are about the same. We are all covered in hair with the exception of only the palms of our hands and soles of our feet. What makes us different to our hairy ape cousins is the evolution of vellus hair, distinct from longer and thicker terminal hair. The bits of your body that you think of as hairless, such as your stomach or the undersides of your arms, are covered in superfine vellus hairs. Each of these is usually only a millimetre or two long and completely colourless. The other type of hair we have, the terminal hair we are more familiar with, grows thicker, pigmented and potentially much, much longer. This is the stuff you possibly have on your head, face, groin, armpits, eyebrows and eyelashes. There are a few differences between some of these types of hair, such as the tendency to be curly and having different sweat glands, but these are superficial when compared to vellus hair. That said, much of the basic biology is the same.

At the base of both terminal and vellus hair follicles there are a number of specialized stem cells. These divide to form a clump at the base or bulb of the follicle, elongate and begin to churn out a protein called keratin, in long continuous strands that form the hair itself. What kind of hair you end up with depends on a variety of factors. The number of stem cells in the bulb of the follicle determines the thickness of the hair and is one of the main differences between terminal and vellus follicles. The shape of the follicle also has a marked effect. A straight and circular follicle predictably produces straight hair, an oval follicle gives wavy hair and for those with tightly curled hair the follicles have a twist along their length that puts the curl into the hair. The overall length of the hair depends on the growth cycle of the follicle. All follicles have a cycle of active growth, termination and rest. The duration of the growth period is what determines hair length. So, on your scalp, once a hair follicle enters a growth phase it can

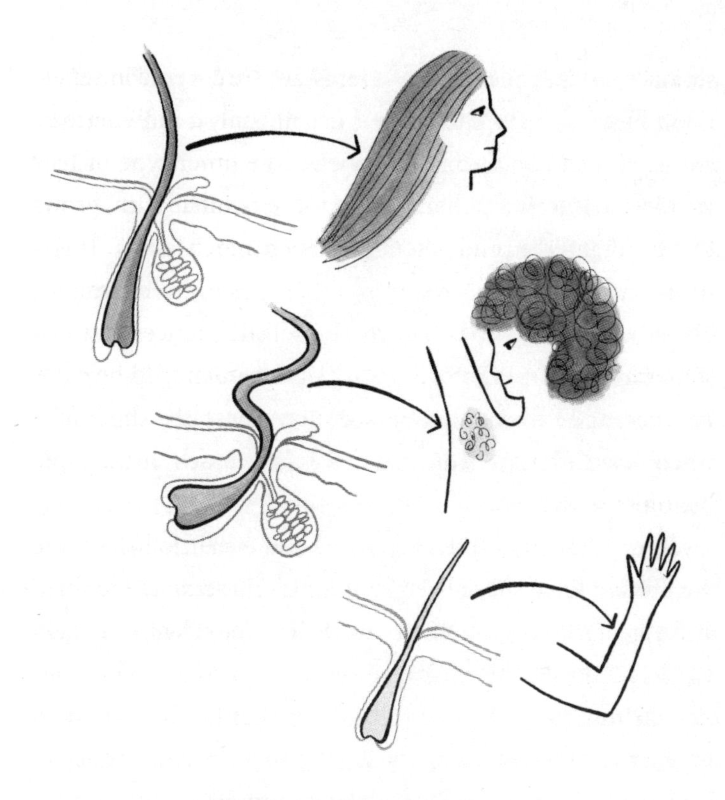

Different types of hair follicle

stay in this phase for between two and eight years, potentially longer. Given that scalp hair grows at about 10 mm a month (⅜ inch) the hair on your head could, given the chance, grow to a metre in length (39 inches). Eyebrows grow slower, at about half the rate of your head hair and only for six weeks, resulting in hairs a little under 1 cm long (⅜ inch). Vellus hair has a very short growth period, far fewer stem cells and a narrower follicle that yields a tiny, straight hair. Following the growing period, the cells producing keratin are shut off from their blood supply, which kills them: the hair is done growing and is itself completely dead, which is when it falls out. The follicle then has a little rest period and starts afresh with a new hair. At least, that is the principle. Clearly for some people there are variations and changes that result in different hair patterns. Classic male baldness is down to a loss of stem cells in the bulb of the hair follicle. What kills these cells off is not the hormone testosterone as you commonly see suggested, but more subtly dihydro-testosterone. Admittedly, this substance is made from testosterone, but you only need a tiny amount to cause the damage. What makes the difference between a bald man and a chap with a full head of hair is not the level of testosterone but the level of the enzyme that converts it to the dihydro version and then how sensitive the stem cells are to this chemical. The effect is to slowly decrease the growing period of the follicle and increase the rest phase. Eventually, the growth dwindles to zero and the stem cells die completely, which is why when it's gone, it's gone and there is nothing you can do to make a bald man's hair follicles restart growth.

The big question for evolutionary biologists, though, is why so much of our hair switched from being of the terminal variety to the vellus type. After all, the hairs on a chimpanzee are all terminal, which is why they are covered in fur. There are a few different ideas about what may have driven this change from our closest evolutionary cousins. An obvious place to start looking for a reason is the differences between the hair on men and women. Men generally have more terminal hair, especially on their face and chest, and when this sort of difference between sexes exists it usually indicates an evolutionary selection based on sexual preference. But all this does is move the question down the line. If being particularly hairy in a man makes the man a more desirable breeding partner, why is that? What is it about being hairy that indicates you would be healthier and your offspring better able to survive than the less hairy chaps?

The most common parasite in human hair is the head louse. This flightless insect is three millimetres long and is adapted to suck the blood from our scalps. Although annoying, it is essentially harmless, unlike its near-identical cousin the body louse, which evolved to cling to vellus hair or clothing and carries diseases such as epidemic typhus and trench fever.

Two possibilities exist. Either having less hair means you have fewer parasites or you are better able to survive in the heat of the midday sun. In the case of the parasite theory we know that creatures like ticks, fleas and lice are far more able to survive in hair than on naked skin. If there is a heavy infestation, these sorts of parasites can also put a significant burden on the body. Losing the hair that allows the parasites to hide on our bodies may have had an evolutionary advantage. The second possibility comes down to controlling body temperature in a hot African climate (see page 9 for more on human evolution). Vellus hairs are particularly good at cooling the body. Sweat produced by glands associated with each hair follicle wicks up onto the tiny vellus hair, evaporates, and in so doing cools the skin. Thick fur can't do this; on the contrary, it provides a layer of trapped air which acts as insulation and traps in heat. We know that *Homo sapiens* evolved under the heat of the African sun and we know we were predominately active during the day, as that is what our eyes are evolved for, so it seems plausible that this need for heat regulation drove us to lose a covering of fur.

The problem with both these arguments is that other primate species have not done the same. If parasites and heat regulation drove us to lose our fur then why did this not happen with chimps, gorillas and the monkeys? There are also the downsides of being less hairy that need to be considered. Vellus hairs may help with heat regulation and parasite load, but at night a covering of terminal hair would keep you warm. The clue may be that this evolution did not happen separately from other traits that humans evolved. As well as our vellus hairs, another unique feature that we do not share with other primates is our intelligence and culture. If this trait developed hand in hand with a reduction of the length and type of our body hair, we could have used our intelligence to develop simple tool use into functional clothing to keep us warm at night. Our culture passed on through language (see page 75) would allow these ideas to spread quickly. It is a rather elegant idea that joins together two strands of our evolutionary history to explain how it is we went from being covered in fur to covered in tiny hairs that make us seem, at least to the casual observer, to be naked apes.

HOW TO FOOL A HUMAN

Adding a dimension

On 27 September 1922 a short film called *The Power of Love* was shown at the Ambassador Hotel Theatre in Los Angeles. The silent film told the story of Southern Californian Don Almeda, his daughter Maria, Don Alvarez – the treacherous man to whom she is betrothed, and the new man in town, Terry O'Neill, that Maria falls in love with. The story itself is deeply unoriginal and sticks closely to the narrative expected by the cinemagoers of this first flourishing of the motion picture. The plot involves lots of deceitful actions by the dastardly Don Alvarez but our heroine Maria, while being wounded accidentally, survives and Terry the hero gets the girl in the end. So, nothing out of the ordinary except for one detail: the 1922 film *The Power of Love* was the first ever to be shown in 3D. Sadly, it is not known if this audience was also the first to suffer from a 3D film headache.

The audience were all issued with special spectacles containing one red lens and one green lens and the film was shot using a camera designed by Harry Fairall and Robert Elder that had two sets of camera lenses side by side about 6 cm apart (2⅓ inches) – the average distance between an adult's pupils. Inside this early film camera were two sets of film spools that recorded the scene simultaneously from two slightly different points of view. The idea behind the technology was that each spool would capture a scene through the equivalent of one eyeball. The trick was then how to deliver the left-hand film to the left eyeballs of the audience and the right-hand film to the right eyeballs. Which is where the red and green viewing spectacles come into play. By projecting the left-hand film through a red projection filter, you end up with a red-tinged film. If you place a red filter, identical to the red projection filter, over the right eye of every member of the audience, the red tinged film won't pass through the spectacle filter and only the left eye will see this version of the film. Repeat the process with the right-hand film, but this time passing it through a green filter and putting a green filter in the left of the audience spectacles, and they only see this right-hand film in their right eye. The result is a three-dimensional experience in which the action seems to move off the flat screen and has depth that comes closer to and further away from the screen.

Three-dimensional images of the non-moving variety were not a new idea. The first 3D-image viewer, or stereoscope, was devised by the prolific inventor and telegraph pioneer Charles Wheatstone in 1838, the year before any practical photographic process existed. Consequently, it was initially used with hand-drawn images, but from the very beginnings of photography people were experimenting with stereoscopic 3D images. Two years later, William Henry Fox Talbot, the inventor of the negative photographic process, approached Wheatstone to begin taking stereoscopic images. The early stereoscopes relied on mirrors, or a partition that you placed your eyes next to in order to allow each eye to see a separate image, taken from a slightly shifted position. Your brain then recombines these into a three-dimensional image, the same way that it recombines images of the real world produced by our left and right eyeballs.

The problem with these first systems is that only one person at a time could view the 3D imagery, which is where the clever red and green filter system, or anaglyph stereoscopy, comes into its own. It was invented in 1853 by a German called Wilhelm Rollmann and so long as you are wearing a pair of the filter-containing spectacles you can see the image. Suddenly, the field of 3D imagery was open to an audience of any size. The process of going from a single still image to a moving one was cracked at the end of the nineteenth century by a Frenchman called Louis Le Prince, so it was no great surprise that 3D eventually made its way onto the silver screen.

In 1922, the premiere of *The Power of Love* won good reviews but the film only ever received one additional screening. Films in 3D remained a minor novelty until 1952, when they burst once more onto the screen. The first of this new era of feature-length films was called *Bwana Devil* and, from the start, a major problem with the technology was encountered. As one critic put it, 'my hangover from it was so painful that I immediately went to see a two-dimensional movie for relief'. The technology seemed to give a large proportion of the audience a headache. This golden era of 3D films lasted just two years and from 1954 to the early 1980s audiences sufficed with two-dimensional films. A brief resurgence in the 1980s brought us classics from some of the then major movie series such as *Jaws 3-D* and *Friday the Thirteenth Part 3: 3D*. While the technology of filming improved a little, with most systems now using polarizing filters rather than colour ones, the basic problem of the 3D headache remained.

At last we can come up to date and the latest revival of 3D films which didn't begin with, but reached a high point with the 2009 film *Avatar*, directed by James Cameron. The film was deliberately shot to be viewed in 3D and proved to be a huge success (until July 2019, when it was overtaken by *Avengers: Endgame*, it held the accolade of being the highest-earning film ever produced, with an income of nearly US $2.8 billion). One of the reasons that the audiences weren't put off by the 3D technology was that we now understood why viewers get 3D headaches and that changed the way James Cameron made the film.

The reason we see three-dimensional objects is because we have two eyes and what is known as binocular vision. Since our eyes are offset left and right by about 6 cm ($2\frac{1}{3}$ inches), the world we see through each eye is slightly different. What each eye perceives on its own is a flat two-dimensional image, detected by the rods and cone cells in the retina at the back of each eyeball. The two images coming from the eyeballs are then interpreted by our brains so that we perceive the world in three dimensions with depth and not just width and height. But, as usual, human biology has an extra wrinkle to contribute and there is more information added to this picture. Whenever you focus your eyes onto an object, two things happen within and between the eyeballs: accommodation and vergence.

To make the light that falls onto the retina at the back of your eye form into a sharp picture, the clear lens that forms your pupil is able to change its shape and thus its ability to focus. This accommodation of the lens ensures that we see things clearly and for those of us who need to wear glasses we are correcting a defect in this ability. Accommodation of the eye's lens is a reflex, meaning that it takes place automatically, without conscious effort on our part, and it goes hand in hand with a second ocular reflex known as vergence.

To see vergence in action you will need to first find a helper or possibly some way to film yourself, as you can't see the effect by yourself. Ask your friend to hold out a finger at arm's length and focus their eyes on the tip of their finger. Now have them slowly move the finger closer and closer

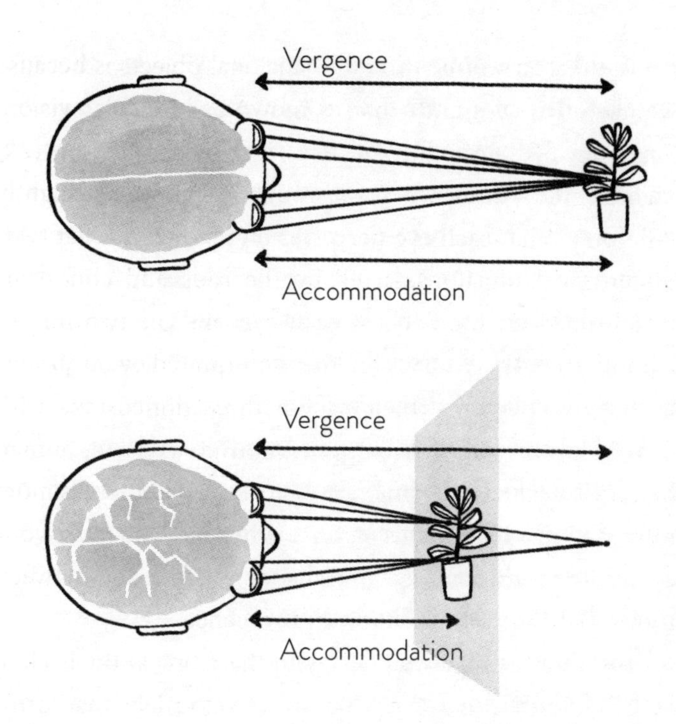

Headache caused by vergence and accommodation mismatch

to their face while still keeping their eyes on the same spot on their finger. In turn, you need to watch their eyes. As the finger moves closer and closer the eyes begin to point inwards towards the nose until your friend is positively cross-eyed. This is vergence, or at least in the case of the finger, demonstration convergence. As an object moves closer to us our eyes cross and converge. If the object moves away our eyes diverge. Taken together, the accommodation-vergence reflex provides our brain with extra information about the three-dimensional world we perceive through our eyes. The more our eyes accommodate and converge, the closer the object is, which provides information that is added to the images we actually see to allow our brains to construct a fully three-dimensional picture of the world around us. Unfortunately, all the clever camera trickery used in 3D cinema with filters and lenses only takes into account one part of this rich information input and ignores the accommodation-vergence reflex.

When you look at a 3D film in the cinema your eyes automatically focus the image on the cinema screen onto the back of your eyeballs. Your eyes accommodate your lenses so that they are focused at a distance appropriate to the distance of the screen. Similarly, in order for both eyeballs to look at the same thing, the vergence of the eyes is adjusted so they point at the exact spot and you are not cross-eyed. The problem comes when the 3D film you are watching is portraying an object that protrudes in front of or behind the screen. Take the instance when the object is in front of the screen. The image each eye receives is creating a 3D image in your brain that should be closer to you than the surface of the screen, consequently your eye lenses should be focused closer and you should have your eyes pointing inwards more. Except the image is still really on the screen. Your brain is receiving contradictory information. On the one hand, it is perceiving a close object, but the accommodation-vergence reflex is saying the object is at the distance of the screen. The feedback from the brain to the eyeball tries to resolve the dilemma, either forcing you to override the accommodation-vergence reflex, in which case the image blurs, doubles and ceases to be 3D, or you ignore the brain and let the reflex win out and see the three-dimensional images. Either way, this constant conflict can lead to eye strain and the typical '3D headache' experienced by so many cinemagoers.

How, then, can you get around this apparently inevitable problem? The key comes to understanding that no matter what is on the film screen at any particular time, almost all of the audience will be looking at the same thing. A good film director will intuitively know where the audience will be looking at any instant, as it is a human trait that we instinctively seek out certain details with our eyes. If a person on screen begins to talk, our point of view will jump to their face, irrespective of what else they are doing. Understanding this is the basis of most close-up magic, along with lots of practice, of course. You can easily test and confirm the theory of where an audience is looking with eyeball-tracking technology. Armed with this information, an astute director of a 3D film will ensure that the point of focus stays in the middle of the depth of field shown. When this happens, the images your eyes receive place the point of focus on the cinema screen at the same point as the accommodation-vergence reflex is trying to force your eyes to focus. Your eyes are now no longer fighting your brain and the headache should – in theory – not start. What this does mean is that directors must avoid the popular habit of the early 3D films where objects were suddenly thrust close to the camera, appearing to leap out from the screen. It is effects like this that cause the most conflict between the brain and yes, cause the headaches. Where *Avatar* differs from hangover-inducing *Bwana Devil* is an understanding of how our brains construct a three-dimensional world view and also how that interacts with the technology of the cinema.

Just as with 3D films, the virtual-reality computer gamer may experience headaches and also motion sickness, particularly if the computer system is not powerful enough. If there is even a tiny delay between head and image movements, a mismatch occurs between the movement part of the brain and the vision part, resulting in a feeling of motion sickness.

Intriguingly, that first 3D film, *The Power of Love* from nearly a hundred years ago, had one other groundbreaking trick up its sleeve that was not taken any further. The directors realized that the technology they were using was essentially displaying two different films at the same time, admittedly showing the same scenes from just 6 cm apart. If they had two different films being watched they surmised that they could produce two alternate endings to the film. As the film came to its climatic conclusion, the audience were told that if they wanted a happy, literally rose-tinted ending to the tale they should only watch through one eye and the red filter. Conversely, if they wanted to see the jealous villain Don Alvarez triumph, then they should watch through the other eye and the green filter. This tragic ending has been lost to history as no copy of the 3D version of *The Power of Love* remains and the 2D version that does exist only has the happy ending. It is an intriguing concept in film-making but was only possible due to the film dating from the era of the silent movie. To create the same effect today, even with the state of the art in 3D film-making, you would need to issue the entire audience not only with proprietary spectacles but also headsets, so the sound heard matched the different endings to the film. All of which has the potential for a whole new level of headaches, not least for the cinema staff.

The art of lying

If you want to know if somebody is lying to you the telltale signs are that they will not maintain eye contact and they will fidget more. On top of that, I'm not good at lying but I am quite good at spotting when others are lying. These are the archetypal responses you will get from a cross section of people from around the globe if you ask them about lie detection. The most commonly believed signs of lying are a lack of eye contact and fidgeting, combined with a belief that you are a poor liar yourself, but good at spotting lies. Perhaps surprisingly, given how common these beliefs are, they are all not quite lies themselves, but at least not true. For those of us who have not had specific training in how to detect lies, our ability to detect untruths is about even. Your chance of correctly spotting if something is a lie or a truth is only 50 per cent, and you would be just as successful by flipping a coin. Equally, almost all of us are no better at telling lies convincingly than we are at spotting them. In work carried out by American psychologists that gathered together hundreds of studies from around the globe, they found that with nearly 25,000 people tested the success rate for identifying either truth or lie was only 54 per cent. If you break it down and look at the success that people had at specifically detecting lies it was even worse than random, at only 47 per cent. Conversely, we are as a species a little bit better at knowing when others are being truthful – 61 per cent of the time we can spot truth.

What, then, of the assertion that liars avoid eye contact and fidget more? The science behind this belief turns out to be just that, a belief and not science at all. This idea and the ideas behind much of lie detection go back to the Italian psychologist Vittorio Benussi, who was born in 1878 in Trieste, which was at the time of his birth part of the Austro-Hungarian empire. Benussi put forward the idea that when a person lies they exhibit increased emotional disturbance. In Benussi's case, he attempted to detect this disturbance through a change in breathing patterns. It was not a very successful experiment but it started the quest for an accurate way to detect the physiological changes caused by the emotional disturbance of lying. In the 1920s an American named William Moulton Marston decided that if breathing rate was not the answer then perhaps tiny changes in blood pressure were. Marston claimed that his lie-detection machine was between 90 and 100 per cent accurate. Subsequent tests by other scientists failed to achieve the dizzy heights of perfect accuracy leading to the conclusion that Marston was perhaps confusing his experimental work with another of his creations, the comic book character Wonder Woman, who famously wielded the magical Lasso of Truth that would compel those caught in its coils to speak only the truth.

Subsequent work on lie-detection machines brought in pulse rates, skin resistivity and measures of sweat production all to create a picture of when lying led to emotional turmoil. The final shape of the lie-detection machine, or polygraph as it had become known, was put together by Leonarde Keeler, a Californian who in 1939 sold his invention to the Federal Bureau of Investigation, the US domestic intelligence and security service. Since then, it has been used all over the world for detecting truth in everything from criminal cases to job interviews. The problem is that, for some time now, many in the scientific lie-detection community have come to the conclusion that the polygraph does not work. At best it may be slightly better than a human without a polygraph, increasing our 50 per cent unaided success rate to 60 per cent. A small improvement but the consequences of the test being wrong can be pretty drastic: a polygraph test could be the reason you fail at a job interview or conceivably result in an incorrect criminal adjudication. Despite this, these tests are still being used as admissible evidence in countries such as Japan and the USA, and in the UK prisoners on probation are required to submit to regular polygraph lie-detection tests. The problem with the polygraph is the assumption made back in 1878 by Benussi that a person lying would be emotionally disturbed. It turns out that even people telling the truth become emotionally disturbed if you hook them up with electrodes and wires to a black box of flashing lights in the full knowledge that the person administering the test thinks you may be lying to them. That anxiety alone gives rise to an increase in heart rate, sweating and breathing rate and thus a false positive result from the polygraph. It is well documented that with appropriate training, or just enough self-belief and fear of failure, a criminal who is lying to avoid prosecution can fool the polygraph. What is more, and this is a fundamental problem with many studies on lying, the test subjects used to check the lie-detection device are invariably students who have been asked to either lie or tell the truth. Even though these students are sometimes incentivized to lie convincingly with a cash reward or the threat of an electric shock, in the end they do not represent a fair test. One of the big issues with any experiment involving human beings is that it is very difficult to test a scientific hypothesis without straying into unethical territory.

The idea that underpins polygraphs – that we are essentially leaking information out through our bodies in the form of blood pressure, where we are looking, or myriad other measurable physiological reactions – has proven to be a hard one to shed for the lie-detection academic community. In the 1970s, Californian scientist Paul Ekman devised and tested a new theory that put forward the idea of micro-expressions. These are facial displays of emotions that last for as little as a twentieth of a second. Humans can't detect these, but using slowed-down film of a subject they become visible and, according to the theory, can reveal when a person speaks the truth. Except that this, just like the polygraph, does not really work either and suffers the same problems of being too easily confused by false positives and being tricked by convincing liars. For any polygraph system to really work you need to first establish a baseline of the test subject's behaviour, such as how often they normally twitch, how much they sweat or which micro-expressions they typically use. You can only do this if the subject can be tested in a completely relaxed and unthreatening environment before you start to search for lies. In a real-life application of a lie detector in a criminal case, for example, this is impossible and the system is doomed to fail.

Despite this failure of the technology, there is some evidence that there still may be a reason to continue using the polygraph, but not because it helps detect lies. Rather it has gained such cultural notoriety that subjects are likely to be susceptible to more recently discovered lie-detection techniques. The latest advances in our understanding of how to spot somebody lying rely on neither physiological nor even visual clues. Rather, it seems that we are most likely to give ourselves away with what we say or write. A recent collaboration between Belgian, Dutch and British scientists analyzed a huge database of emails made up of conversations between businesses bidding for awards. In the 8,000 or so different interactions they looked at the language used by the participants and checked the veracity of how each company described itself. They found that when an email was not telling the whole truth about the company bidding for the award there were a few distinct telltale signs. The author would use fewer personal pronouns such as 'I' and 'you' in order, the researchers theorized, to distance the email's author from the lie. In addition, the lying emails were less self-deprecating and used more flowery language with unnecessary and repetitive descriptions. With hindsight, neither of these findings is perhaps that surprising but what is novel was the discovery that in longer email threads, lying email writers would very quickly mimic the linguistic style of the person they were trying to convince with their lies, possibly to ingratiate themselves and make themselves seem more genuine. Based on this analysis, the research team devised a computer algorithm that could analyze an email exchange and predict with some 70 per cent accuracy who was and who was not lying, much better than the usual 50 per cent accuracy an unaided person can achieve.

Backing up this idea that visual and physiological signs are of no use in determining truth is work that came recently out of Canada. In this study female test subjects were shown a film of a woman who had been asked to look after a stranger's bag while the owner had to go elsewhere for a few minutes. The film came in two versions, one in which the woman watching the stranger's things goes into the bag and steals an item. In the other version, the woman patiently keeps watch on the bag until the stranger returns. Half of the subjects saw the theft and the other half did not. The test subjects were then asked to answer a simple question but they were all told to answer it the same way. They were asked if the woman stole anything from the bag and told to answer that she did not steal anything. So, half the test subjects had to lie while the other half told the truth. All of these responses were filmed and the results played back to a large group of volunteers who had to determine who was telling the truth and who was lying. But there was an added wrinkle to this research. Some of the test subjects, as they were being filmed either lying or telling the truth about the bag theft, were asked to wear a niqab, a type of face veil worn by some Muslim women that covers the entire face except the eyes. The impetus for this particular addition came after judges in the USA, Canada and the UK ruled that witnesses may not wear a niqab when giving testimony, in the case of the UK ruling specifically because the judge felt that the jury must be able to see the witness's full face to be able to gauge if they are believable and presumably being truthful. The results of the study showed that if the test subject was wearing a niqab the volunteers were more likely to correctly identify when they were lying and when they were being truthful. Rather than hindering our ability to detect lies from truth, being able to see a person's face makes it harder for us.

All the evidence to date makes it abundantly clear that we are, on average, rubbish at detecting lies. Even with clever gadgets and detection devices we can still only improve our success rate a little bit above a completely random chance. However, there are ways of working out if a person is lying. Firstly, concentrate on what they say and not how they look. If possible, remove the physical person completely from the equation and examine what they say in written form, as this allows you to concentrate on the words and not be distracted by other false cues. The real key, though, is in what you ask, and understanding that it is an easier task to tell the truth than it is to lie. In the case of an interview that is part of a criminal investigation, a person telling the truth must remember just what happened and recount these details. A lying person, though, must invent a story and carefully make sure that they are constantly getting fabricated facts to line up and correlate. This is a harder mental task and has what the researchers call a greater cognitive load. Once you realize this, the trick is to pile more things onto the interview subject's cognitive load. A standard tactic is to ask them to relate the timeline of the incident in reverse chronological order. For the speaker of truth this is a little more difficult, but for the person lying it's much harder and thus they are likely to make a mistake. A method used when trying to establish the true ideological viewpoint of a suspect, often those accused of terrorist offences, is to ask them to defend the mainstream standpoint and then to play devil's advocate and describe the opposing view. No matter how well prepared you are, your ability to put forward an ideology you believe in is much, much easier than to present the opposite view. Consequently, your true allegiance will become apparent. The aim of both of these tactics is to make it harder for the lying person to concoct and maintain a coherent story, hoping that they will slip up, say something contradictory and give the person conducting the interview a way to lever open their falsehoods. It's a powerful technique that can result in much increased detection of lies, in excess of 80 per cent accuracy. But since you probably don't find yourself interrogating a person very often it's not exactly something you can use in a day-to-day setting.

Animals are good at lying, too. The North American killdeer plover feigns a broken wing to distract predators, and Arctic foxes lie to their own pups, making false warning cries so the adults can get a share of food. Koko the gorilla used sign-language to tell her most egregious lie, when she broke her enclosure sink and blamed one of her keepers.

You can, however, turn the idea of emotional disturbance on its head, specifically the idea about liars not holding eye contact. When you are in a conversation it is a perfectly natural response to look away when trying to recall some information. Looking somebody in the eye is quite distracting and if you need to concentrate on remembering some details of an event you will normally look away momentarily. If you are trying to establish if somebody is lying, get them to relate the details of the incident, with as much information as possible, while holding your gaze and not looking away. It's not that you will see in their eyes some telltale sign of lying, but instead you are putting more cognitive load onto them. Listen to them carefully, don't worry about anything but what they say and if they are lying they are much more likely to slip up and get their story wrong. That's how you spot a lie.

The power of nothing

When I was growing up in South London it was a treat when my father would take me into the heart of the capital city at the weekend, usually on a Sunday. We would invariably go to explore either the Science Museum or the Natural History Museum often followed by a visit to a restaurant on Gerrard Street, the heart of London's Chinatown. Both activities have had a lasting impact on me, endearing a love in the first case of science and Chinese cuisine in the second. I remember that some of the restaurants in Chinatown used to advertize prominently that they did not use MSG and I recall as an inquisitive child asking my father what it meant. This was when I became aware of a thing called Chinese Restaurant Syndrome that was something to do with the use of a chemical called MSG that was added to the dishes in Chinese cooking. As a science geek I knew that MSG stood for monosodium glutamate and that it was a flavour enhancer, but beyond this I had no idea what this syndrome was supposed to do and had never personally felt any ill effects from eating in Chinatown, except possibly after eating more of the delicious food than was entirely needed. I further recall that my aunt did suffer from this syndrome and consequently avoided restaurants where MSG was used.

Subsequently, as my knowledge of science grew, the gaps in my understanding of this strange syndrome were filled in. The chemical monosodium glutamate is just a version of glutamic acid that has been first dissolved in water, making it into glutamate, and then crystallized by adding sodium. These building blocks are both found all over the place in biology. The first, glutamic acid, is one of the essential building blocks of proteins, and sodium is a vital element responsible for a huge array of biological functions such as nerve transmission and thus ultimately how your brain works. Given the ubiquitous nature of the constituents and that when monosodium glutamate is added to water it immediately splits into these two parts, it is hard to see how it could cause the headaches, numbness, dizziness, palpitations and shortness of breath associated with the syndrome. I'm not the only person to wonder about this. There have been numerous and repeated studies using double-blinded protocols (see page 110) trying to replicate the effect and in every case the results were negative. So, what does this mean? The science tells me that monosodium glutamate does not cause the effects it is blamed for. Yet people will swear that it does. Not only that, and this is the crucial bit, they do display the symptoms of the syndrome that can be measured and quantified – just not when it's done as part of a blinded scientific trial. We are entering the murky world of the placebo, and its evil twin the nocebo, effect.

The word placebo comes from the Latin and means 'I shall please', while nocebo means 'I shall harm'. The idea of a placebo has been around for a long time, with the nocebo being relatively new. Medics in the sixteenth century readily admit in various treatise that in some cases a sugar pill is the appropriate treatment. The first instance of research being carried out on the placebo effect was by a British physician called Dr John Haygarth. In 1799, he performed a small study using so-called Perkin's Tractors, very expensive metallic spikes imported from the USA that supposedly helped cure all manner of ailments when placed gently on the affected area. Haygarth showed that the same effect could be achieved using a simple whittled stick of wood.

Since then, the placebo effect has become an accepted part of medicine, although it is still mostly not understood. Initially it was considered to be just a nuisance, as when conducting clinical trials it could mess up your results. If, for example, you are trying to establish how well a new drug treatment works, the standard trial would be to treat some patients with the drug and give other patients no treatment. You then compare the patients treated to those untreated to see what the drug can do, except that if you do find an effect this may just be a placebo effect. The drug may in fact be useless, and to control for this you now need to treat a third group of patients with sugar pills, but tell them it is the real thing and compare these placebo controls to the real drug. It all adds to the complexity of real-life clinical trials. We now know that for certain problems the placebo effect is as good as any drug and worth using as an actual treatment.

Take irritable bowel syndrome as an example. This is an unpleasant condition whereby the patient suffers from a combination of symptoms like abdominal pain, bloating, flatulence, constipation or diarrhoea. It has been found that placebo pills or sham treatments have a 40 to 50 per cent rate of success in sufferers. Your initial thought may be that the syndrome itself is not real and thus not surprising that it is cured by a not-real treatment. However, the symptoms of irritable bowel syndrome are very real, measurable, and for sufferers a genuine problem. It is what is known as a set of functional symptoms which means that while there is no cause for the symptoms at the site of the complaint, the brain really does feel the effects and the body responds appropriately. You can have a huge range of functional symptoms from epileptic fits, extreme pain, loss of limb function and even blindness. In each case the effect seems to be a neurological one where the brain and the tissue affected are not communicating correctly. The symptoms are real, just not caused the way you might expect. With irritable bowel syndrome this may explain why a placebo works so well as it helps reset the communication by using the brain's own expectations on itself. It also turns out that even if you know you are being given a placebo you may still get a placebo effect and you do not need to be deceived into thinking you are being given a real drug. When Professor Ted Kaptchuk at the Harvard Medical School in the USA carried out a test of placebos for irritable bowel syndrome in 2017, he carefully made sure that the patients knew exactly what they were being given as part of the trial, notably sugar pills with no drug in them. Even so, the placebo effect kicked in and some patients reported being cured of the syndrome after years of struggling with endless failed treatments. One patient for whom the placebo worked was so desperate to continue on the fake medication that they pleaded to be given further prescriptions of the sugar pills in the full knowledge that this is what they were.

While this wrinkle in the placebo effect is fascinating, the experiment was carried out because working with placebos throws up some serious ethical issues. If a doctor tells a patient that they are being given a series of pills to take that contain a drug that will help cure their condition, the patient then agrees to this and accepts what the doctor has told them. If all along the doctor was giving a placebo, not only have they just lied to the patient potentially ruining the trust the patient had for them, but they are also breaking the patient's consent which was for the active drug and not the placebo. Hence the ethical dilemma many doctors worry about. What

Kaptchuk showed was that, in some cases at least, the doctor can be completely open and honest with the patient, tell them the pills they are prescribing are placebos that have no drug in them and still get the placebo effect, thus getting around the ethical dilemma.

The placebo effect is not just the prescribing of sugar pills. Surgical placebo has also been shown now, in many instances. In the case of knee surgery on patients with arthritis, the standard procedure involves inserting a special device into the knee joint to remove debris and clean the rough wear from the surface of the knee bones. The process involves a general anaesthetic and then a small incision made on one side for the instrument to be inserted. A study was done by a group based in Houston in the USA where some patients who underwent this treatment went through the whole process of anaesthesia and the incision being made, but then nothing else. The incision was stitched up, they were allowed to recover and told the operation was a success. Many of the patients reported that the pain they had previously felt had gone away. The placebo surgery was a success. The two issues that arose from this study were first the ethical angle of pretending a patient has undergone surgery and secondly that the placebo surgery had nearly as good a result as the actual surgery.

Placebo also has a downside, which is when it becomes known as nocebo. This was first observed when patients were specifically told about all the negative side effects of drugs they were being given. If you are told you may suffer a side effect, you are more likely to suffer the side effect. What is more, you might get side effects when you are prescribed a placebo drug. For example, when researchers retrospectively looked at the data from clinical trials on beta blockers used to lower blood pressure, they found that patients in the control groups who received placebo treatment reported the same level of side effects as those who were given the actual beta blocker drug. Since each patient does not know if they are being given a real drug or a sugar pill and it didn't make any difference what they were given, they all reported the same levels of side effects, there is a good chance that all the side effects are due to the nocebo effect after been told to expect side effects.

It is clear that the placebo and nocebo effects are real. People start out with a set of symptoms that are quantifiable, measurable and not just in their heads. They undergo a sham treatment, in some cases in the full knowledge of what is happening, and the end result is a change in their symptoms, getting better with placebo and worse with nocebo. The results can be impressive, especially if the patient is suffering from functional complaints where the problem is really to do with how the brain and nervous system are communicating. Worryingly for some clinicians, the placebo effect can be as great if not greater than the use of the drug under test, raising the question of whether the drug is doing anything at all and if all its effects are placebo. Controversial work by an American psychologist, Irving Kirsch, pulled together a whole range of studies on antidepressants that seem to show that they are only slightly better than placebo. There are some extra complications, as the mere act of being involved in an elaborate drug trial is a form of talking therapy and this may be what was helping the patients recover from their depression. Then again, it could just be about how you measure how poorly a person is with depression and what constitutes an improvement. Another psychiatrist, James Warner from Imperial College in London, showed the opposite effect, that antidepressants were twice as good as placebo. The debate among scientists is currently deep in the thick of how experiments and analyses are carried out. None of this, though, tells us how or why placebo works.

An interesting possibility comes from work on Siberian dwarf hamsters and the work of Pete Trimmer at the University of Bristol. This particular variety of hamster is a tiny little furry ball that will sit quite comfortably in the palm of your hand. Unlike their bigger golden or Syrian hamster cousins they are a grey colour and have a dark stripe along their backs from nose to tail. You can find them for sale in most pet shops as they make excellent pets for young children for two reasons. Firstly, they are tiny and thus take up minimal space and, secondly, they only live for about a year to eighteen months which given the usual length of time a child takes an interest in a pet, about four months, only leaves the parents looking after the animal for twelve. And these are the same reasons the animals are used in behavioural studies in laboratories. For some time, it had been known that the immune response mounted to an infection in the Siberian dwarf hamster depended on the lighting conditions it was subjected to. If the lights above the animal's cage were set to mimic a Siberian summer, with long periods of light and brief episodes of dark, then the immune response was vigorous and strong. However, if the lighting was set to winter, the immune response was essentially turned off. Trimmer took these observations and created a computer simulation that predicted the cost-benefits of such a strategy and in so doing provided convincing evidence for why the hamsters did this. Key to this is understanding that the immune system is a resource intensive part of the body to activate, especially for a tiny creature struggling to survive a harsh Siberian winter. Consequently, it appears that the tiny hamster has evolved to only mount an immune response to disease when it knows it has the plentiful resources of summer available to it. An infection in winter may overwhelm the hamster and cause its death, but so too would mounting an immune defence against this infection. Instead, the hamster gambles on leaving the infection to, hopefully, abate with a minimal immune response, saves itself precious resources and survives through the winter. This raises an interesting possibility. What if this seasonal immune response based on environmental situation were present in other animals, possibly even humans? It is not too far-fetched to imagine that in our distant evolutionary past *Homo sapiens* also had a similar seasonal immune response. We would rarely see it today in action as, since the advent of farming in the Neolithic period, scarcity of resources in winter have been drastically reduced. It is possible that we have been left with the evolutionary quirk of an immune system that can be boosted when we think we are in ideal environmental conditions, such as, for instance, when we have just been handed a course of pills we are told will help cure us, even if the pills are just sugar.

While this is an intriguing possibility that may go a little way to explain why the placebo effect exists, it does not help us understand how it works or how we can exploit it. The answer to that may come once again from Harvard Medical School, where the genetics of the placebo effect are being looked into. One of the perplexing things about the placebo effect is that it can be very dependent on the patient, with some showing strong effects and some none at all. Which is what started Kathryn Hall looking at the genetics of the effect to see if there were specific genes that could be linked with a strong placebo effect. So far, eleven genes have been found to correlate with being responsive to placebo. It was already known that people who rated as open to new experiences and who were outgoing or extroverted had better placebo responses but for the first time the genetics have been opened up. One gene in particular stood out as interesting. The gene that encodes for a protein called catechol-O-methyltransferase or COMT seemed to be particularly associated with placebo. This protein is part of the breakdown process of dopamine, a chemical used in brain nerve transmissions specifically with feeling pleasure, reward and pain-relief. There is a mutated version of the COMT gene that is quite common in the Caucasian population. If you have two versions of this gene you end up with much higher levels of dopamine in your brain. It has been seen that patients who had this double mutation used fewer self-administered painkillers when they were in hospital, presumably because they maintained a higher dopamine level. Hall set up a study to look at COMT mutations using the condition of choice for placebo trials, irritable bowel syndrome. Patients with the double mutation were nearly twice as likely to respond to the placebo. The exciting possibility is that by looking at the genetics it may be possible to tailor the treatment a patient receives.

The legendary effects of monosodium glutamate felt by those eating Chinese food are very real to those who suffer from the syndrome, but we can probably put the causes down to nocebo, coupled with overeating. In the same way, the modern equivalent of the Chinese Restaurant Syndrome, gluten intolerance, has also been shown to be similarly difficult to replicate and is thus probably also mostly down to the nocebo effect. Ironically, the best way to potentially cure this, especially in those genetically susceptible, will be to administer sugar pills and a course of placebo.

Fooling your mouth

In order to convincingly make food from one source that pretends to be something completely different, you need to jump some considerable hurdles. Scientists around the world are starting to turn their attention to the problem for a simple reason. It is the proverbial elephant in the room when it comes to farming and nutrition: we eat too much meat. The production of flesh for consumption by humans is, without any doubt, the least efficient way to make food and energy for us to eat. This is not a matter of ethics or politics, although those play a huge part, but one of thermodynamics and scientific laws. Any biological process is inherently somewhat inefficient. There is a loss of energy as it moves through the various processes and organisms. With very few exceptions, all biology on the planet derives its energy from the sun. Plants use photosynthesis to capture this energy, wasting a little as they do, and turn the energy into complex molecules like sugars, carbohydrates and proteins. Animal life on the planet then eats the plants as an energy source or, if you are higher up the food chain, you eat the animals that ate the plants. At each step energy is lost, and it is not just a few percentage points of loss, but in the order of a half or three-quarters. So, if we choose to eat meat, be it fish or chicken or beef, there will have been a big loss of energy between the original plants that captured sunlight and the meat we eat. Conversely, if we go direct to the plants and skip the intervening animal, we bypass the inefficiency. Compared to meat, plant-based food takes less land, less water, fewer resources and has a lower carbon footprint to make an equivalent number of food calories or kilojoules. So, we need to eat less meat in order to feed a booming world population, but people like to eat meat. It is a high-status food in most societies and lots of us think it tastes great. In

China, as income levels grew between 1970 and 2007, meat consumption per person went up by 400 per cent.

Which is why making meat from plants is on the agenda, but it turns out to be really difficult. This is not because of anything inherent with meat, but just that what makes one food different to another is a subtle blend of sensory inputs. Many of these are relatively easy to get right, others less so.

As an example, consider a beefburger. It may seem that the most obvious thing to get right is the taste, but this turns out to be relatively straightforward. When acclaimed biochemist Pat Brown was approaching retirement age, he decided that he could possibly rest on his laurels a little. He had been the inventor of several significant breakthroughs in molecular biology, had a slew of awards and was recognized as one of the USA's top scientists. Concluding that food production was one of the top environmental problems we faced, he decided to turn his attention to the issue of making a meat substitute from plants. He founded a company called Impossible Foods and ten years later you can now buy the Impossible Burger. Getting the taste right was apparently relatively easy; after all, there are only a handful of different tastes or chemicals you can detect on your tongue. Trickier was to nail down the smell which is so vital to our perception of how a food tastes. Brown's team used the best kit they had access to, and given they were working at Stanford University, one of the top universities in the USA, they had some fancy kit. They found that much of our perception of flavour, both the umami taste and the smell, comes from the organic compound called haem. We are all familiar with haem as it is found in haemoglobin, the protein in our red blood cells that allows oxygen to be transported around the body. It is what gives blood its characteristic colour and meat much of its flavour. It is also found in many plants in a somewhat different form. After some experimentation, the Stanford group genetically engineered yeast cells to produce a plant-based, haem-containing compound, which when added to their fake burger made it taste and smell incredibly meaty.

Lab-grown meat offers a different approach to making a beefburger without having to use meat from animals. Cells from a cow are grown in highly specialized laboratory conditions, harvested and then packed into burger shapes for consumption. Since the first lab-grown burgers in 2013 received high praise from food critics, companies around the world have created lab-grown chicken, fish and even duck.

Getting the texture of the fake minced beef to feel right in the mouth is the really difficult bit. This is where the vast panoply of tricks developed by food scientists came into play. The basic texture is recreated with vegetable protein extracted from wheat and soya. The addition of plant gums and various clever types of starch allow the mix to take on a distinctive chewiness and resistance to your teeth. Getting the feeling of juiciness just right needed the addition of neutral-flavoured coconut oil that behaves much like the saturated fat found in meat. The latest version of the Impossible Burger is, according to all reviews, a pretty convincing fake. Some may be happy with this, but according to Pat Brown this isn't good enough. If the aim of the meat-free burger is to convince people to eat less meat, we need to change habits and the only way that will happen for those that are not already ethically motivated is if the fake burger is better than the real thing. Impossible Foods are aiming to make a product that tastes better, is juicer, more consistent, cheaper, easier to cook and better for not just the environment but the consumer as well. It is a lofty goal but could make a difference if they achieve it.

Of course, you may be asking, why go to all the effort? The real issue is that people just prefer to eat meat. But that is probably more to do with cultural choice than taste choice. Not only that, there are plenty of non-meat products that are in the same ballpark as real meat, mostly from Asian sources. Tofu many people are familiar with, but there are also products such as mock chicken, made from layered and fermented skin taken during the making of tofu. Another popular meat-like substitute is seitan, produced by washing the starch out of wheat-flour dough until all that is left is the gluten protein network. The result is a chewy, protein-rich block that does not taste of a huge amount on its own and is not far removed from chicken. The bottom line for vegetarians, though, is that they rarely want to eat something that is spookily like meat. It may be that if we want to make more efficient use of our food production capabilities, we need to stop eating meat and just start learning to like vegetables instead.

SURVIVING IN A CROWD

Three's a crowd

Human beings are, with few exceptions, inherently social creatures. No matter what you may think about your antisocial relatives or colleagues, we are all driven to interact with our fellow humans. It is an evolutionary imperative that we find ourselves gathering into social groups and spending time with each other. Long-term isolation from other people is not good for us and can lead to permanent psychological and mental harm. Solitary confinement is regularly and possibly unethically used as a form of punishment in many judicial systems. The point is that we need other people around us. Over the course of history this has led to the growth of settlements from small family units to disparate tribes and on to huge gatherings in cities of millions of individuals. With these larger and larger gatherings come the chance for us to congregate in larger and larger crowds.

What exactly constitutes a crowd is worth a little thought. There is an English aphorism that while two is company, three is a crowd, but I doubt anybody would agree with this literally. If you see three people standing at the bus stop you don't think of it as a crowd. It's a bit of a conundrum and technically a form of sorites paradox. This idea was first put forward as the paradox of the heap by Eubulides of Miletus, an ancient Greek philosopher from the fourth century BCE. The paradox of the heap goes like this. I put one million grains of sand on the floor and we can all agree it is a heap. We can also agree that the removal of a single grain is not going to change it from being a heap. However, if I keep removing one grain we will eventually get to a single grain of sand left on the floor and this is clearly not a heap. So, when did that happen? When did the heap stop being a heap? The answer, especially when applied to people in a crowd, is that you know it when you see it, and specifically when you are a part of the crowd.

Psychologists have defined three zones of personal space. For partners and our children is 'intimate space', extending to half a metre (1.5 ft); 'personal space' – 1.5 m (4 ft) – is for friends and family; and 'social space' is up to 3.5 m (12 ft). If people encroach into an inappropriate space, it is deeply uncomfortable and will unconsciously change your behaviour.

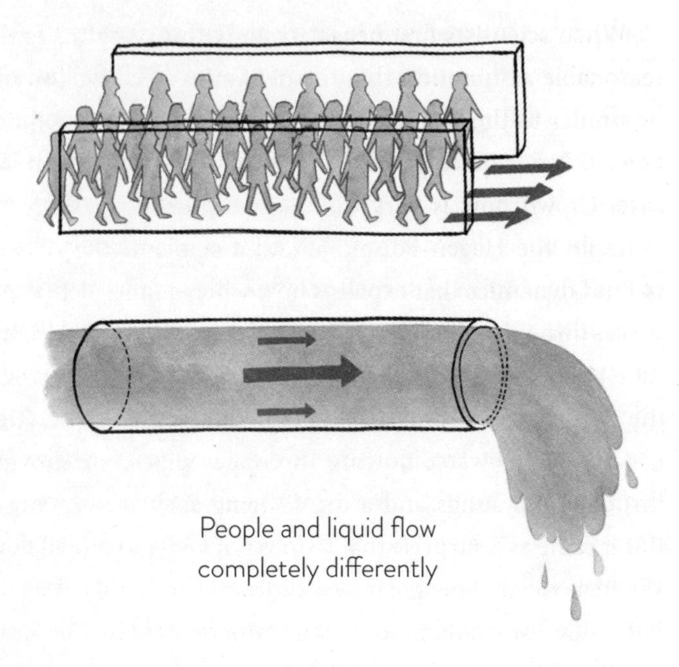

People and liquid flow completely differently

What we can say is that the industry standard for the safe upper limit in a moving crowd is four people per square metre (about four people per ten square feet). It's quite hard to envision what this amounts to, so imagine drawing a square on the floor 50 cm by 50 cm (about 20 inches by 20 inches) and then standing and walking in a space this large when surrounded by hundreds of other people. The space between the front of you and the back of the next person is a scant 20 cm (8 inches) and from your shoulder to your neighbour's shoulder only 5 cm (2 inches). Four people per square metre is uncomfortably tight.

When scientists first began to study this, they made the reasonable assumption that the dynamics of crowds would be similar to the various equations that defined how fluids flowed. But it quickly became apparent that this is not the case. Crowd flow is very different to fluid flow. Take, for example, the Hagen-Poiseuille law, a staple of the science of fluid dynamics that explains how when a fluid of any sort passes through a long tube it flows fastest in the middle of the tube and slowest near the tube walls. The friction between the static tube wall and the moving fluid slows it down. This law applies to water flowing in copper pipes, air moving through your lungs and a drink being sucked up a straw. But it came as a surprise that it does not apply to crowd flow. When a crowd of people moves down a corridor, the Hagen-Poiseuille law would predict that the people in the middle of the corridor move fastest and the ones along the wall slower. It turns out that there is no friction or even something analogous to friction between a person and the wall they are walking next to. Consequently, a crowd moving through a corridor all move at the same rate, irrespective of how close they are to the walls. People are in the habit of breaking the laws of fluid dynamics.

The egg-timer enigma

Once it became apparent that large groups of people don't obey the usual laws of fluid dynamics, scientists working on crowds started to look out for other peculiar behaviour. One such oddity was observed very directly by Professor Keith Still of Manchester Metropolitan University when he found himself squashed together with some 75,000 other people trying to get out of Wembley Stadium in London. The date was 20 April 1992 and we can be sure of that as the event was the Freddie Mercury Tribute Concert for AIDS Awareness. At the end of the concert, Professor Still found himself part of the massed crowd trying to get back out of the stadium through one of the many exits. As he made his way to one of these exits he became stuck right in front of the exit moving forward at a painfully slow pace. The crowd became almost stationary and he remained there for over half an hour, inching towards the exit. Once the initial frustration of being delayed had passed, his analytical scientific brain kicked in and he began to wonder why the flow through the exit was so slow. According to Hagen-Poiseuille, the part of the crowd directly in front of the exit should move fastest, but he was in this part of the crowd and not moving. By the same token, the crowd along the edges should move slowest but when he looked he realized that the people near the edges were the ones who were moving. For a normal fluid flow, in front of the exit should be fastest and along the edges the slowest. You can see this in action if you watch an old-fashioned sand egg timer. A dimple forms directly above the narrow hole as the sand in the middle flows fastest. At the edges where glass meets sand, the sand barely moves. The dynamics of the crowd of Freddie Mercury fans back in 1992 was the exact opposite predicted by the Hagen-Poiseuille law. Once out of the stadium and back at his desk, Professor Still began to work out what was going on and why crowds didn't behave like regular fluids. He called it the egg-timer enigma.

What became apparent is that you need to start thinking about crowd flow in different terms. Within liquids such as water, the interactions between the liquid and the vessel or pipe holding the liquid are really important, at least when compared to interactions between individual molecules of the liquid itself. The molecules in water, for example, are free to move about each other but the tenacity with which they cling to the walls of a pipe slow them down, hence the Hagen-Poiseuille law. Within crowds of people, the

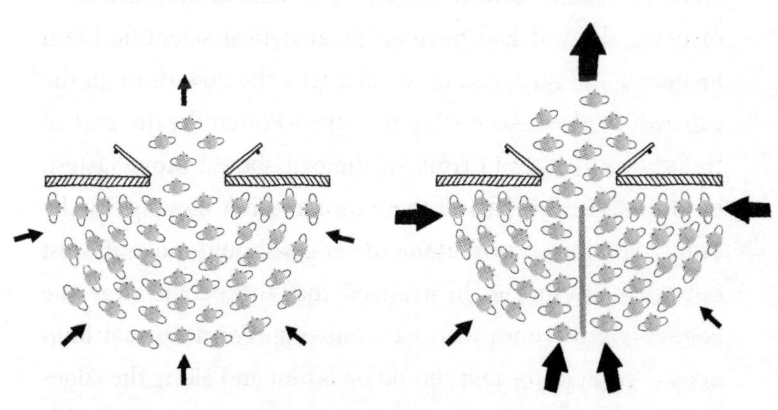

A central barrier speeds up a crowd through a doorway

opposite is true. If you consider the individual molecules of a crowd, by which I mean people, and how they interact, the rules change. I am quite happy to brush up against a wall, press on an inanimate barrier or squeeze around a corner. So long as I have room to walk next to a wall, there is no appreciable friction between me and the wall and I can walk just as fast as if I was not next to the wall. On the other hand, it's the other people moving in a crowd that slow me down. We all have a built-in personal space around us and when this space is invaded, especially by strangers, we feel uncomfortable. Possibly more important than this, we can empathize with our fellow crowd members and avoid invading each other's personal space. When you walk in a crowd you are constantly monitoring your own personal space and that of those around you, trying not to bump into each other and to maintain an acceptable distance. All of which makes us slow down. The important friction within a crowd is not between the fluid of people and the walls containing them, but between individual molecules of this people fluid. Which is why we humans break the Hagen-Poiseuille law.

When a crowd of people try to get through a small space, such as an exit from a stadium, a couple of things happen. The people directly in front of the exit move forward and as the density of the crowd squeezes tighter, the space per person decreases and the number of interactions between people goes up. As this increases we slow down trying to avoid an embarrassing bump or treading on each other's toes.

Conversely, the people right along the edges of the stadium have a wall of some sort on one side that has no personal space. These people need to worry about fewer interactions with other people as one side is people free. Consequently, they are not slowed down and can continue to walk at a reasonable pace. The egg-timer flow in a crowd is reversed and those nearest the edges move fastest, which explains Professor Still's stadium problem. It also leads to a couple of interesting implications.

> Whoever invented the sand timer is a mystery. The Babylonians and Egyptians had water clocks, which are similar, but replacing the water with sand allowed for much longer and more accurate timers. We do know that the first properly documented sand timer appears in an Italian fresco painted in 1338. For such a fundamental invention its history is surprisingly unclear.

The first is that if you find yourself in a crowd trying to move through a small gap or restriction of any sort, your best bet to minimize the time it takes you to get through the exit is to work your way to the edge. You can then scoot along the wall or barrier and should beat the people in the middle of the mass. I have tried this myself and, based on my anecdotal evidence, it seems to help. It is also the best way to board a train, especially busy Tube trains in places like Tokyo and London. Don't stand in front of the door waiting for it to open, but instead just to the side. This situation is further complicated by the flow of people out of the Tube train carriage increasing person-to-person interactions.

The really clever application of this observation is used in crowd-control situations where lots of people need to get through a small space. The solution is to add a barrier at right angles to the door or exit, but directly in front of it (see diagram on page 183). The barrier pokes out from in front of the exit space and would seem to be blocking it. When you first see one of these barriers in place, your gut reaction is to wonder which idiot put that there. Our common sense would clearly tell us that such a barrier is cluttering up the exit and bound to lead to delays getting a crowd through the space. As with so much science, common sense is often not a great way to judge what works. The interposing barrier in front of the exit effectively splits the crowd into two groups, one on the left and one on the right. It creates an additional wall directly in front of the exit and people next to this thus experience fewer person-to-person interactions. Just like the members of the crowd along the edges, the people next to the barrier move faster. The crowd as a whole passes through the exit space quicker than when the barrier is not there. Seeing it in action, as I have had the delight to do so with Professor Still, is almost magical, as your preconceived common sense is blown away by a demonstrable bit of science magic. All of which goes back to having the right person, Professor Still, in the wrong place, a stationary crowd at Wembley Stadium, at the right time, 20 April 1992.

Step locking and the Millennium Bridge

When crowds of people do get under way and start to move they have been found to display some other peculiar effects, especially on the world around us. On 10 June 2000, the London Millennium Footbridge was opened to the public. It's a beautiful steel span that was originally and appropriately called the Blade of Light. On the opening day, the public flocked to it and it was crossed some 90,000 times, with up to 2,000 people on it at any instant. But there was a problem. It wasn't that the bridge couldn't cope with the numbers, but something unexpected happened. The bridge began to wobble and sway in quite an alarming fashion. At its worst, it was swaying, left and right, by 10 cm on the central section. The design firm, called Arup, were needless to say extremely worried and quickly instituted a system to limit the number of people on the bridge. The media went crazy for the story as the bridge design was already controversial. It had been judged by some to be ugly and too expensive at over £18 million. Not only that, but it opened some £2.2 million over budget and two months late. It was of course extremely embarrassing for Arup when, after just two days of limited access during which the bridge wobbled so violently the authorities were concerned people would fall and be injured, they decided to close the bridge for safety reasons. The press howled and fingers of blame were pointed. Journalists took delight in explaining that the cause of the wobble was obviously people walking in step. Not only that, but the Albert Bridge over the River Thames just upstream suffered a similar but more minor issue. It even had 100-year-old notices on it stating that 'All troops must break step when marching over this bridge' to prevent it from trembling. While it is true that the Millennium Bridge wobbled because the people were walking in step, what made the people walk in step was new to the engineers.

When you put your foot down, the force of the impact is sent into the ground. Most of this force goes vertically into the ground but there is some that goes horizontally, pushing the surface you are walking on not just forwards, but also sideways. It's the sideways bit that is the hardest to get your head around and it is this that made the bridge wobble. Since human beings have two legs, our centre of gravity does not sit directly over either leg. Consequently, when you take a step the force goes from the centre of gravity, about where your belly button is, at a slight angle down to your foot. Since there is a slight angle, there is a small element pushing the surface you are walking on to the side and away from you. It is only a small amount and if you imagine a crowd of people walking along a bridge, the assumption had always been that we were

Step locking on a bridge

all walking at a random pace out of sync with each other. All those tiny sideways forces would be happening at different times and half would be to the left while half to the right, so they would cancel each other out.

So far so good, but the Millennium Bridge was a cutting-edge design never before built that took bridge technology to new places. What makes this particular bridge so beautiful, in my opinion, is the delicateness of its design. They called it a Blade of Light for a good reason as it is a very, very slim and insignificant structure. It is also, technically, a suspension bridge. The common image of a suspension bridge is something like the Golden Gate Bridge or even Tower Bridge in London, both of which prominently feature huge tall towers and an enormous cable going up and down over the towers. The Millennium Bridge does not at first sight have these features. What the designers did was shorten the towers right down to mere stubs and erect them not vertically but at about forty-five degrees to each side. The cable is there, but when you are on the bridge it does not impede the view and almost seems like a bit of decoration. But it is this cable that is holding the whole thing up. The consequence of the design means that the cable has to be very, very highly tensioned. This in turn means that the bridge is a bit like a stretched rubber band and prone to vibrations.

The classic example of a wobbly bridge was the Tacoma Narrows suspension bridge in Washington State, USA, which opened in July 1940. On 1 November that same year, in a strong wind, the bridge reached its resonant frequency, wobbling until it violently and spectacularly collapsed. The only casualty was a cocker spaniel named Tubby, trapped in a car on the bridge.

Everything in the world has what is known as a resonant frequency, which is the rate at which it naturally wants to vibrate, and we use tuned resonant frequencies on musical instruments such as guitars to make music. Because it is so highly tensioned, the resonant frequency of the Millennium Bridge is about one vibration per second. This just happens to be exactly the same as the average rate of human steps. We take about one step a second, and your right foot hits the ground about once every two seconds.

Now, Arup knew this and thus knew that people on the bridge would be walking at a multiple of the resonant frequency of the bridge but, since the acknowledged wisdom was that all the little sideways forces from the steps would cancel out, it wasn't a problem. Except it was.

It turned out that we humans are exquisitely sensitive to movements of the surface we walk on, far more so than people imagined. If the surface under your feet is moving even a few millimetres from left to right, you unconsciously adjust your pace so that when the floor shifts to the right, you put your right foot down. If you don't do this, and put your right foot down as the floor moves left, the moving floor shunts the top of your body to the right, unbalances you, and you stumble. So, as soon as the bridge began to wobble by even a millimetre or two, everybody on the bridge locked step with the bridge. No longer were all the steps randomly cancelling each other out. To make matters worse, when you try walking on a horizontally moving surface, to make yourself more stable, humans adopt a wider gait. You start to walk with your feet further apart, which means that each step pushes sideways more than usual. These two effects combined meant that when the critical number of 160 people were walking on any one span of the Millennium Bridge, it began to wobble left and right. Add another twenty or so people and that wobble would grow, and grow, and grow – until the bridge was throwing people about and it became almost too difficult to move.

So, the Millennium Bridge was wobbling because of people locking step but what nobody could have anticipated was how sensitive humans were to locking step with a moving floor.

Solving the problem of the Millennium Bridge was not difficult: all that was needed was some damping devices fitted to quash any wobble that may develop. The challenge was adding these without ruining the beautiful and graceful design of the bridge. It took Arup less than two years to work out a solution and then install all the dampers at a further cost of £5 million. The bridge reopened in January 2002 and these days is a fantastic, and rock solid, un-wobbling way to cross the Thames with glorious views in all directions. However, it is to this day known by the locals, with some affection, as the Wobbly Bridge.

Your queue will always be slower

If people walking along a bridge can reveal new science, it turns out that even the humble queue can prove scientifically surprising. A classic example of this is the observation that when you are in a queue of people, the adjacent lines of people also queueing always seem to go faster than you. You have probably dismissed this as paranoia. However, don't disregard this observation so quickly, as it turns out to be often true. Most of the time the other queues really are going faster but it's not paranoia, just mathematics.

Imagine you have been to the supermarket and are waiting in line to pay for your goods. To the left and right of you are identical length queues. What are the chances that you will be in a slow queue and one of the other queues will beat you to the cashier? This boils down to a question of mathematical probability: what is the chance that one of the other two queues will be faster than your queue? When you have questions of probability cases like this it is often easier to first work out the answer to the opposite situation, that is, what is the chance that your queue is the fastest?

Let us assume that there really is no conspiracy theory against you and the cashiers are not trying to make you wait longer than anyone else. Instead we need to presume that each cashier is equally fast at processing the payments at the till and the size and complexity of each customer's transaction is randomly determined. Or to put it another way, you didn't join the queue with the cashier on their first day in the job and the people in front of you are not all pushing mountainous trolleys full of goods. Given these fair assumptions, then, it should be an even split on which queue is fastest: each of the three queues has a one third or 33 per cent chance of being fastest. So, if your queue is 33 per cent likely to be the fastest, that means the chance of one of the other queues being faster is the inverse of this, a two thirds or 66 per cent chance.

All of which means that when you find yourself standing in a set of queue lines, you are twice as likely to find that one of the other queues is faster than your queue (a 33 per cent chance times two is 66 per cent). So, it's just basic probability that makes you wait. It should be noted that if you want to improve your chances of being in the fastest queue, then go to the line at the edge. Since you only have one other neighbouring queue, there is now a 50 per cent chance you will be the fastest.

I should also point out that if you are lucky enough to be in the queue that is winning the speed race then psychologically you don't mark it up as a negative experience. Consequently, you often don't register the event. When you then find yourself next in a queue but this time a slow queue losing the race, you only recall the times other queues went faster and not the times when you were in the fastest queue. It can all start to seem like the queues are out to get you and paranoia creeps in.

This is just the tip of the iceberg of what is known as queuing theory, a field of mathematics that looks at how to optimize lines of human beings waiting for a service. The theory dates all the way back to 1909, when Danish engineer Agner Erlang was trying to work out how to reduce waiting times for people to be connected on a telephone network. Once he had solved this problem he realized that it could be applied to any queuing situation. Subsequent mathematicians have worked out a way of expressing any queuing behaviour with a code of just two letters and one number. The first letter of the code tells you how often customers arrive at the end of your queue. If they show up at a fixed, or deterministic, interval, the first letter of the code is a D. If they show up at a random interval that has a peak of probability in the middle, technically a Markov process with a Poisson distribution, then the first letter is M. The second letter in the code refers to how long it takes each person in the queue to be dealt with by a sever. Again, this could be a fixed length of time (code D) or a random distribution (code M). Finally, the last character is the number of servers that are dealing with the people in the queue.

Given this way of thinking about queues you can see that the simplest type of queue would be D/D/1. The first D means that the time it takes for a new customer to arrive at the back of the queue is fixed and always the same. Similarly, the second D tells us that the time it takes to deal with each customer is also fixed. This situation is almost unthinkable if humans are involved and a D/D/1 queuing system is more likely to be encountered in a factory, for example a machine putting lids on jars. A queue will not form in this example if the arrival rate of the un-lidded jars is less than the rate at which the single, lid-attaching machine can screw on a lid. The mathematical formula for waiting time on a D/D/1 queue is very simple.

However, the maths behind the queue in the supermarket is different. In this case each line of people represents an M/M/1 queue as the customers turn up at random and also take a random amount of time to be dealt with. These random elements mean that sometimes there will be no queue and then, without warning, the queue length will grow out of control and, without the intervention of bringing in more cashiers, can quickly gridlock the shop. The mathematics are complicated and, just as with other crowd behaviour, no longer intuitive. For example, if you double the time the cashier takes to deal with a customer the average wait time doesn't also double, instead it quadruples, increasing by a factor of two multiplied by itself.

Interestingly, this type of queuing also describes what happens in public lavatories. I am sure you will have noticed that the queue for the female lavatory is always much, much bigger than the male one. At the bottom of this, pun intended, is a fundamental issue, pun also intended. According to a global study, the average time taken by a bloke in a public loo is thirty-nine seconds compared to eighty-nine seconds for women. So, it's a little bit more than double the time. If you draw a parallel with my supermarket queue example, this time is analogous to the time the cashier takes to deal with each customer. Since this is an M/M/1 queue this leads to women waiting at least four times, probably five times as long as men. On top of this, despite planners allocating more space for female toilets, the male lavatory often has the same number or even more individual facilities, since urinals take up much less space.

In the 1850s, public lavatories for women were a much-debated issue of female emancipation. There were no Victorian public lavatories for women, despite being plenty for men, so women were unable to venture far from home. Only with pressure from groups such as the Ladies Sanitary Association did women break free from what had become known as their urinary leash.

The solution can be found once again in the mathematics for an M/M/1 queue. To reduce the wait time in the queue for women's toilets so that it equals that for men you need to at least double the number of facilities. If the men's loo has two toilets and four urinals (a total of six facilities), the women should be allocated twelve or more toilets. Sadly, this rarely happens, since a urinal takes up about half the space of a toilet cubicle, and assuming the ratio of two urinals to each cubicle in the men's loo, the women should be allocated three times as much floor space for their twice-as-many facilities. The exponential maths rarely get taken into account. Which should give you something to think about next time you are waiting in a queue.

The best way to board

While waiting for a queue to the toilet may be a bit of a nuisance, some human crowd behaviour can have significant financial implications. At the end of 2018, All Nippon Airways brought in a relatively small change to their systems that has the potential for saving the company billions of yen each year. All they did was change the way they boarded passengers onto their planes.

The All Nippon Airways company has the largest fleet of aircraft in Japan. While it is not the official national airline – that's Japan Airlines – it has at the time of writing 232 aircraft, which operate predominantly within Japan but also internationally. It currently manages nearly a thousand take-offs and landings each day, but from a financial perspective aircraft only make money when they are in the air moving passengers or freight about. The bit where the plane is on the ground taxiing, being refuelled, being cleaned and moving passengers on to and off the plane is not profitable. In fact, it is quite the reverse. The estimate for average ground fees charged to an airline by an airport is about US $35 for every minute the plane is on the tarmac. Which may not seem that much, but in the case of All Nippon Airways their aircraft make over 350,000 visits to airports each year. The cost quickly mounts up. Consequently, it has been a topic of some considerable academic and industrial research on how to shorten the period on the ground at the airport. There are many things that are a fixed duration that cannot be changed, such as the time it takes to trundle the plane to and from the landing strip. In addition, some aspects of the turnaround time can be run in parallel at the same time. The moment the plane lands the disembarkation of passengers takes place simultaneously with refuelling, cleaning and the unloading and loading of baggage. One of the big time-costs for any airline is the time it takes to get all the passengers onto the plane, have them stow their hand luggage and take their seats. It is something that airlines can take control of themselves and involves an understanding of how humans behave in crowded confined spaces.

The standard boarding system used by many airlines around the world is known in the trade as block boarding. It involves calling people to the airport departure gate starting with those whose seats are at the back of the plane. Once that block is boarded you shift forward in the aircraft cabin and often three separate blocks are called, one after the other. There are some subtle variations to this that have been tried, from boarding blocks alternating left and right to boarding window seats first, followed by a block for middle seat passengers and ending with aisle seats. All of these systems yield a boarding time for an aircraft such as a Boeing 757 of about thirty minutes. For reference, the Boeing 757 usually only boards through its front door and seats about 130 passengers with three rows on each side of the single central aisle. This figure also assumes that passengers are bringing on just one piece of hand luggage and one personal item at most. Small changes in boarding time can be made, but what came as a shock to the traditional airlines was so called random boarding.

The low cost or no-frills airlines that began operating at the start of the 2000s experimented with doing away with reserved seating, which had been part of air travel since its beginning. One of the consequences of this is that if you don't have reserved seats, you may as well just let the passengers pick their own at random. This free-for-all system where the passengers are just let on and allowed to sort it out themselves can create an unpleasant rush for what people perceive as the best seats. However, it often outperforms the standard block-boarding method by up to five minutes. It would appear that random boarding creates a feeling of urgency in us as we know that if we don't get a move on we may end up in an undesirable seat. Tighter restrictions on hand luggage may also be factor, as it is the stowing of hand luggage that slows the whole boarding process down.

Academics began to take a closer look at what people were doing as they boarded a plane, the sort of interactions that took place and the different strategies each of us used to get to our seat. It became apparent that a relatively small number of the passengers can have a big effect on the overall boarding time. If those passengers slower to take their seats board early the whole process is delayed, whereas if they board last it makes no difference. Equally, groups of people can actually speed up the boarding process as they effectively act as a perfectly coordinated mini-block that are all sitting together, unless they have lots of hand luggage. While collecting data on the current boarding systems was easy enough, getting companies to commit to experimental boarding regimes was much harder. Fortunately, with all the data collected on boarding it became possible to make computer models so that experiments could be tried in an entirely virtual environment.

A number of second-generation systems began to appear based on these virtual tests. One of them was proposed in 2008 by Jason Steffen, then working as an astrophysicist at the Fermilab near Chicago in the USA. Professor Steffen has informed me that it was at Seattle airport in 2006, while he was stuck waiting in the access corridor that led from the gate to the aircraft, that he began to wonder what was causing the delay. He could see a reason for delay at the gate or in passport control or even security, but not in the access corridor. So, as he put it, he decided to allow himself a brief distraction from his hunt for planets around distant suns and went to work on aircraft boarding. The system he came up with was created using an optimization algorithm. Starting with a random passenger order, he had a computer model the boarding time, record the result then randomly swap two passengers and calculate the new boarding time. He then took the passenger order that gave the quicker result and tried another random swap, and kept repeating this process. After some 10,000 boarding simulations the computer model had settled on an optimum arrangement. It's a little bit complicated, but starts by boarding passengers seated in window seats on every other row on the right of the aircraft. Then come the matching window seat passengers on the left, followed by the remaining window seats on the right and then back to window seats on the left. At this point all the passengers by the windows are seated and you start on the middle rows and end with the aisle seats. The results are impressive and can, in theory, nearly halve the time taken to board the passengers: that's down to fifteen minutes for a Boeing 757. I've seen the Steffen method in action in a test I was involved in running and it was indeed impressively fast. We used a mock-up aircraft half the length of a Boeing 757. The Steffen method chopped about a third from our boarding time compared to block boarding. But there was a problem, and it came outside the aircraft, as we had to get the sixty-six test passengers lined up in the correct and rather convoluted order. This part of the process took the best part of half an hour. While it definitely gets people on the aircraft quicker, an extra half-hour waiting at the gate getting in order is not going to be a winner for the airline customer service record and no airline has ever used the Steffen method to date. But that doesn't mean they won't in the future.

A group of researchers from Naples in Italy took the boarding process to the next, digitally customized level. They realized that no matter what system you used it was the random interactions between people that caused the delays. In particular, passengers with lots of hand luggage or who were less mobile caused more delays. The research team

The Steffen boarding method

designed a system that customizes seat allocation based on the individual and can be used for boarding planes where the passengers don't have pre-allocated seats. It uses digital cameras to scan passengers as they walk through a single-file entryway leading to the departure gate. Using this scan, the system gives each passenger two ratings: one determined by the bulkiness of their hand luggage, the other a mobility rating. The algorithm behind this system then allocates a seat based on these two ratings and a knowledge of who is immediately ahead of the passenger. When the passenger gets to their seat they find the aisle empty and they can take their seat quickly and without hassle. The algorithm is underpinned by the Steffen method but adds in a cunning way of gathering extra information and then of dispensing it. It is very much a work in progress, but could cut the boarding time for a Boeing 757 to just twelve minutes.

All these ingenious ways of dealing with a flow of humans are still a way off. However, given that there are huge financial savings to be made for the airlines, there is a definite hunger for new ideas in the industry, as demonstrated by All Nippon Airways. The new system they are using is called a reverse pyramid and is really a simplified version of the Steffen method layered onto the basic block-boarding method. Tests show that you get an average ten-minute saving on boarding time. For All Nippon Airways with their 1,000 flights a day, this equates to a yearly saving of at least 10 billion yen (US $125 million or £100 million). When those sort of savings are possible, change is likely to happen.

Beware the phantom traffic jams

There is one last example of human crowd behaviour that once you know the science behind it will drive you crazy. We have all experienced it and there is nothing you can do to stop it happening to you. You find yourself zooming along a motorway, or depending on where you are it may be a freeway, a highway, an expressway or even, for some reason, a parkway. Suddenly, the traffic ahead slows down and you come to a complete standstill. Over the next ten to fifteen minutes you slowly crawl forward, only for the traffic to just as suddenly start to speed up. There is no traffic accident or roadworks to explain the short, sharp traffic jam. What you have just experienced is a phantom traffic jam.

The science behind traffic congestion throws up some interesting observations but the first time phantom jams were observed in an experimental situation was in 2007. The experiment was set up as part of a Japanese television programme by researchers at Nagoya University. Professor Yuki Sugiyama arranged for twenty-two cars to drive around and around in a circle about 73 m across (240 ft). The experiment took place at the Nakanihon Automotive College just outside Nagoya, where presumably it was easy to find lots of people willing to drive the cars. The drivers were told to try to maintain a speed of 30 km per hour (19 miles per hour) and due to the size of the circular track only had a gap of 10 m from the front of their own car to the one in front (33 ft). Given the speed and proximity

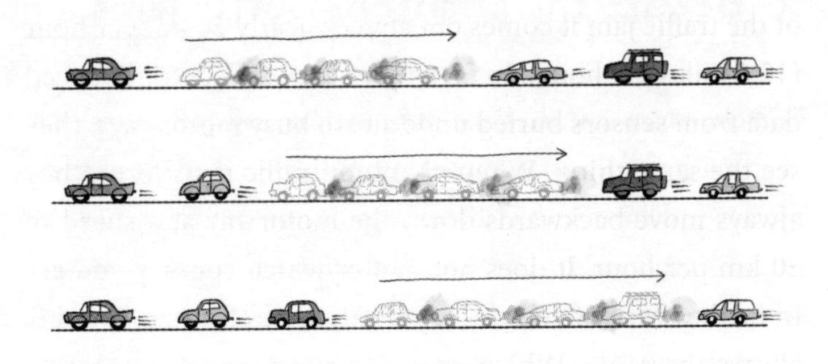

Phantom traffic jam moving backwards

of the next car it made for a hair-raising drive. The video of the experiment was widely circulated on the media and can still be found online. Initially, the cars trundle around the track quite happily, but then something odd happens. A cluster of cars forms on the track right in front of the camera position which quickly becomes a slow-moving phantom traffic mini-jam. In a version of this experiment carried out in the UK that I had a chance to watch, we saw the same jam form, but in our case the cars came to a complete stop.

What becomes apparent from the footage of these experiments is that the traffic jam is itself not stationary. The clump of slow or stopped cars moves backwards around the circular track. By which I don't mean that the cars go backwards, just the location of the traffic jam. Cars drive away from the front and simultaneously pile up at the back. When you measure the speed of this backwards movement of the traffic jam it comes out at very nearly 20 km per hour (12.5 miles per hour). In fact, when they collect traffic speed data from sensors buried underneath busy motorways, they see the same thing. When phantom traffic jams form, they always move backwards down the motorway at a speed of 20 km per hour. It does not matter which country you are in or the types of vehicles involved, the traffic jam speed is always the same. Which on the surface looks like an odd coincidence, until, that is, you try to dig into the causes of phantom traffic jams.

Traffic researchers initially used fluid flow equations to model the flow of vehicles, but just as with crowds of humans moving through small gaps it turns out that we don't behave the way physicists would like us to. Subsequent studies and research have observed that there are three types or phases of traffic flow on a motorway. First you have what is known as free flow, where cars travel at a variety of speeds and the speed of each car is determined solely by the driver and their desire to adhere to speed limits. As the traffic density increases, you get to a point where suddenly you shift to synchronized flow characterized by the traffic all driving at the same, or very nearly the same speed. In this situation, it does not matter what speed you want to drive at: you are limited by the other cars on the road and often the difference between lanes is only a few kilometres or miles per hour. The final phase is known as the wide-moving jam, which is basically the same as a phantom traffic jam. The root cause of changes between these three phases would appear to be just an increase in traffic density, or the number of cars and lorries squeezed into each kilometre or mile of road. A sudden increase in density can cause a shift from one phase to the next.

The cause, then, of a phantom traffic jam is not always some dramatic event, such as a near-miss collision, or sudden swerving of a car. While these sudden events usually do cause a phantom jam, they are fairly rare occurrences and phantom traffic jams far more common. Instead, when the traffic density is high enough, an accumulation of small but normal human reactions can lead to each car in the line having to reduce speed and that sets off a chain reaction. The result is that cars down the line become slower and slower until one reaches a standstill and you have initiated a phantom traffic jam that then propagates backwards down the road. The most obvious of these normal reactions is that when a car in front of you brakes a little to slow down you are almost inevitably going to brake a little bit more than them. It's a natural desire to not want to crash into the back of the vehicle in front. But there are some more subtle causes like what happens when you change lane on a motorway.

Consider for a moment that when you change lane you are occupying two lanes simultaneously during your manoeuvre. Just because your vehicle is only half occupying each lane that still counts as two lanes since a half-width lane is of no use to other vehicles. If you occupy two lanes for that moment you are now effectively two vehicles on the road and thus increase the traffic density. Another common density increase happens when you reach a junction and more cars enter the traffic flow, often accompanied by lots of lane changing to exacerbate the problem even more. Any of these events can initiate a phantom traffic jam.

Traffic jams are not just mathematical but also economic problems. According to capitalism, goods are allocated through ability to pay or a first come, first served basis. Typically, users take advantage of roads until they are full and a traffic jam occurs. Economics implies that jams are inevitable unless you change to a payment system and put tolls on roads.

What I find intriguing, though, is the constancy of the speed of the phantom traffic jam backwards down the road, irrespective of cultural differences in driving habits or local traffic regulations. The fact that the speed of the jam is always 20 km per hour (12.5 miles per hour) means that the root cause of the jam is human biology. In the end, the reason behind the formation of phantom traffic jams is down to human reaction times, human perception ability and human psychology. It has nothing to do with the vehicle you drive or the place you drive it.

It is possible to prevent phantom traffic jams from propagating down the road. If you leave a bigger gap between yourself and the car in front you have more time to react if they brake. Consequently, you over-brake less and match your change in speed closer to theirs. Reducing the overall speed of traffic when it is in synchronized flow has the same effect and this is why so-called smart roads now have variable speed limit signs. As the traffic density goes up and the flow switches to synchronized, the speed limit is reduced and it helps prevent the formation of a traffic jam. Clearly, not changing lane when you are in synchronized traffic flow also helps, and again you see this suggested on overhead signs on smart roads. The thing that rankles, though, is that no matter what you do it can only ever help people behind you on the motorway. If you find yourself in a phantom traffic jam, no amount of careful driving can speed up your exit from the jam. The sorts of behaviour I described above – leaving bigger gaps, slowing down and not changing lane – are purely altruistic. It's a form of prisoner's dilemma: if we all do these things then the traffic flow is faster for us all. On the other hand, if you act selfishly then the traffic slows, and while you go faster than the other people on the now-jammed-up road you won't go as fast as if we all acted altruistically. So, the next time you are stuck in a phantom traffic jam, while you can enjoy the intriguing science behind your situation you are helpless to do anything and that's why it drives me crazy.

ACKNOWLEDGEMENTS

The business of writing books is, quite frankly, beyond me, which is why I am so grateful to my agent Sara Cameron and all the staff at Take 3 Management for navigating the process on my behalf. Then there is the issue of making sure the words I put down on the computer are in a sensible order. Once again, this is thanks to the lovely folk at Michael O'Mara Books. In particular, I am indebted to my editor Gabby Nemeth, who not only tidies it all up and takes the extra commas out, but who also helped spawn the idea of this book over a lunch in a cafe in Clapham, South London. If I recall, it was the story of the London Millennium Bridge and step locking that piqued her curiosity.

Finally, this book would not be here without the help of my family, who put up with me being at home, and especially my wife, who provides so much of the inspiration, even if she does not realize it.

INDEX

Page numbers in *italic* are illustrations.

H

I

K

L